Water Security for Better Lives

This work is published on the responsibility of the Secretary-General of the OECD. The opinions expressed and arguments employed herein do not necessarily reflect the official views of the Organisation or of the governments of its member countries.

This document and any map included herein are without prejudice to the status of or sovereignty over any territory, to the delimitation of international frontiers and boundaries and to the name of any territory, city or area.

Please cite this publication as:
OECD (2013), *Water Security for Better Lives*, OECD Studies on Water, OECD Publishing.
http://dx.doi.org/10.1787/9789264202405-en

ISBN 978-92-64-20239-9 (print)
ISBN 978-92-64-20240-5 (PDF)

Series: OECD Studies on Water
ISSN 2224-5073 (print)
ISSN 2224-5081 (online)

IWA Publishing
ISBN: 9781780405766 (Paperback)
ISBN: 9781780405773 (eBook)

The statistical data for Israel are supplied by and under the responsibility of the relevant Israeli authorities. The use of such data by the OECD is without prejudice to the status of the Golan Heights, East Jerusalem and Israeli settlements in the West Bank under the terms of international law.

Photo credits: Cover © radiusimages/Inmagine LTD, © Taro Yamada/Corbis.

Preface

Water security is one of the defining challenges of our time. By the middle of the next century, over 40% of the global population will live under severe water stress. As global population increases, so will tensions among different water uses.

This challenging outlook on water security, together with an increased severity in floods and droughts brought about by climate change, is an urgent call for better managing water risks, including water shortages, excesses, pollution, and other risks to freshwater systems (rivers, lakes, aquifers). The key lies in adopting an approach based on knowing, targeting, and managing water risks.

A water supply crisis – a decline in the quality and quantity of freshwater – is perceived by many experts to be one of the top five global risks, both in terms of likelihood and impact. Until recently, water risk management has largely focused on providing an appropriate short-term crisis response, aimed at protecting human lives and critical assets from disasters, ignoring the long-term management of water security.

This report, Water Security for Better Lives, proposes a fundamental shift in the approach followed by governments to tackle water security challenges. It argues in favour of a risk-based approach to improve water security in a cost-effective manner. Water management involves decisions about the allocation of risk reduction efforts and their associated costs. The usefulness of a risk-based approach lies in making these judgements explicit, through both informed policy discussion with relevant stakeholders, as well as policy responses tailored to the agreed levels of risk.

Implementing a risk-based approach relies on better understanding the context of water supply, demand, quantity and quality; in correctly assessing the relevance of economic efficiency versus equity concerns; and in identifying policy and economic instruments that promote greater water security. The aim is to develop strategies that avoid, reduce, transfer or improve our capacity to bear water risks.

The report also provides guidance on how to implement this approach from a government perspective and addresses the need for policy coherence when managing the trade-offs between water security and other economic, environmental, and social policy objectives.

I am delighted that the OECD is joining forces with other international organisations, governments, business and civil society to address the water security challenge and promote better water policies for better lives.

Angel Gurría
OECD Secretary-General

Table of contents

Tables

Figures

Acknowledgements

This report brings together OECD work on water security from the last two years that has been undertaken under the umbrella of the OECD Horizontal Water Programme (HWP). The HWP brings together expertise from the Environment Directorate, the Trade and Agriculture Directorate, the Public Governance and Territorial Development Directorate, the Directorate for Financial and Enterprise Affairs, and the Development Co-operation Directorate.

Preparation of this report was co-ordinated by Gérard Bonnis from the Environment Directorate, under the overall supervision of Anthony Cox, Co-ordinator of the OECD Horizontal Water Programme. Substantive inputs to the report were provided by: Kathleen Dominique from the Environment Directorate (Chapter 1 and Annex A); Ian Prosser, Science Director, Water for a Healthy Country Flagship, The Commonwealth Scientific and Industrial Research Organisation (CSIRO) Canberra ACT, Australia (Chapter 2); Quentin Grafton, Director, Centre for Water Economics, Environment and Policy (CWEEP), Crawford School of Economics and Government, The Australian National University, Canberra, Australia (Chapter 3 and Annex B); Suzi Kerr, Senior Fellow, Motu Economic and Public Policy Research, Wellington, New Zealand (Chapter 4); and Tamás Medovarszki from the Environment Directorate (Annex C).

The report also benefited from the presentations and discussions at the OECD Expert Workshop: Water Security: Managing Risks and Trade-offs in Selected River Basins, held in Paris on 1 June 2012. The country case studies from the workshop can be found at *www.oecd.org/water*.

Editorial support by Janine Treves and Patrick Love is gratefully acknowledged. Sama Al Taher Cucci assisted in the preparation of the report for publishing.

Abbreviations

2030 WRG	2030 Water Resources Group
ADB	Asian Development Bank
AQUASTAT	Information system on water and agriculture (FAO)
AUD	Australian dollar
BoD	Burden of disease
BRIICS	Brazil, Russia, India, Indonesia, China and South Africa
CatNat	Insurance Scheme for Natural Catastrophes (France)
CEWH	Commonwealth Environmental Water Holder (Australia)
COAG	Council of Australian Governments (Australia)
CSIRO	Commonwealth Scientific and Industrial Research Organisation (Australia)
ETS	Emission trading scheme
EU	European Union
EUR	euro
FAO	Food and Agriculture Organisation (UN)
GBP	United Kingdom pound
GDP	Gross Domestic Product
GHG	Greenhouse gas
GL	GigaLitre (billion Litres)
IBT	Increasing block tariff
IEA	International Energy Agency
IPCC	Intergovernmental Panel on Climate Change
IWMI	International Water Management Institute
IWRM	Integrated water resource management
MDB	Murray-Darling Basin
MDGs	Millennium Development Goals (UN)
MENA	Middle East and North Africa
MINAS	Nitrogen and phosphorus accounting system (Netherlands)
NPS	Non-point source
PES	Payments for ecosystem services
UN	United Nations
UNEP	United Nations Environment Programme
USD	United States dollar
USGS	United States Geological Survey
WATSAN	Water supply and sanitation
WEF	World Economic Forum
WFD	Water Framework Directive (EU)
WHO	World Health Organisation (UN)
WWAP	World Water Assessment Programme (UNESCO)

Executive summary

Water security is a major policy challenge confronting governments around the world. In the absence of significant reforms of water and water-related policies, the outlook for water is pessimistic. Water security in many regions will continue to deteriorate due to increasing water demand, water stress and water pollution. Governments need to speed up efforts to enhance efficiency and effectiveness in water management to better manage the risks of potential water shortages (including droughts), water excess (including floods), inadequate water quality, as well as the risk of undermining the resilience of freshwater systems (rivers, lakes, aquifers). By taking a broad, long-term vision that emphasises the explicit management of water-related risks and trade-offs between these risks, governments are more likely to meet their water-related economic, environmental and social objectives.

A risk-based approach addresses water security first and foremost by determining acceptable levels of different risks in terms of the likelihood that they will occur and the potential economic or other impacts if they do, and balancing this against the expected benefits of improving water security. While it is generally too expensive, and often technically impossible, to fully eliminate water-related risks, a risk-based approach can help to ensure that the implicit level of risk implied by different policy actions reflects societal values. For example, a number of cities worldwide – including London, Shanghai and Amsterdam – have protection against flood events of a magnitude that are only expected to occur on average once in 1 000 years, while New York planning has only protected the city against a 1-in-100 year event. Following the 2013 Sandy storm, New York is now considering how to strengthen its flood defences further.

A risk-based approach is also flexible, and the accepted level of risk can be adjusted at relatively short notice should more cost-effective measures to mitigate the risks become available, or if new opportunities for economic development warrant action to further reduce the level of risk. For example, a new housing or industrial development may justify increasing flood defenses for a neighbouring river, which may not have been justified if the land was used for agriculture or a natural park.

In practice, however, it is often natural disasters – and not new opportunities – that prompt countries to revisit the acceptable levels of water risks implicit in their policies and measures. For example, countries often revisit flood defense standards following a hurricane or major storm, or address water shortage challenges during or following a major drought. A risk-based approach triggers a move from reactive to more proactive policies. Instead of responding to water crises, which can often entail excessive costs to society, governments can establish a process to carefully assess and manage the risks in advance and to review these on a regular basis.

By identifying water-related risks, and helping actors agree on acceptable levels for these risks, a risk-based approach can facilitate the process of allocating water risks between uses. For example, there are many regions where available water resources have been over-allocated and a more complete understanding of the risks and trade-offs around alternative uses of water can help to identify the benefits and policy options for improving the allocation of water between agriculture, urban, industrial and ecosystem users. This does, of course, raise significant political economy questions.

Once set, the acceptable levels of water risks should be achieved at least possible cost. Economic instruments, such as charging appropriately for water use and pollution, can help achieve this. Water pricing has been critical in decoupling water use from continued economic growth in almost one-third of OECD countries in recent decades. Introducing prices that reflect water scarcity can help reduce demand to levels that can avoid the premature construction of new water supply infrastructure. In Sydney, Australia, for example, analysis shows that if scarcity pricing had been introduced at an appropriate time it could have reduced water demand to a level which no longer required the development of a costly new desalination plant.

Setting acceptable levels of water risk should be the result of well-informed policy choices and trade-offs with other related (sometimes conflicting) security objectives – e.g. food, energy, climate, biodiversity. This is because policy measures aimed at security or other policy objectives in one area may result in spill-overs in another: efforts to increase energy security and reduce greenhouse gas emissions through biofuel production, for example, can result in reduced water or food security, while objectives to enhance food security can lead to overuse of pesticides and fertilizers, contributing to water pollution. More coherent policy approaches are increasingly being applied in a growing number of countries. For example, shifting agricultural support from direct production and input support to payments that are decoupled or even support environmental objectives has reduced incentives to intensify and extend production, thereby helping to improve water resource use efficiency and lower water pollution from agriculture.

Water security is about learning to live with an acceptable level of water risk. This requires a better understanding of the risks, ensuring that the level of risk that is used for planning and policy purposes takes account of social preferences, and managing risks and trade-offs between risks and across water and other policy objectives at least cost to society. The key success factors are to know, target and manage the water risks:

- **Know the risk.** Identify water-related risks, the likelihood and potential impact of their occurrence, how people perceive them, and make sure stakeholders have the information they need to understand and address different kinds of water risks.

- **Target the risk.** Consider whether the additional benefits of improved water security warrant the additional costs to society of achieving these improvements, and set levels of water risk accordingly. Policy objectives other than water security (for example food security, energy security, climate security, protecting nature) and the interrelated nature of water risks should be considered when weighing the benefits and potential costs to society of a given level of water risk.

- **Manage the risk.** Implement a policy mix to reduce hazards and limit exposure and vulnerability in order to achieve acceptable levels of risk at the least possible economic cost. Economic instruments can play an important role, as they can fundamentally alter the incentives facing water users, provide explicit signals about the likelihood and potential cost of water risks, and provide financing to support actions to offset risks. Managing water risks also require a coherent approach between water policies and sectoral and environmental policies.

Chapter 1

Why does water security matter?

Water security is about managing water risks, including risks of water shortage, excess, pollution, and risks of undermining the resilience of freshwater systems. This chapter provides the rationale and conceptual basis for a risk-based approach to water security. It argues that a risk-based approach has many advantages over current policies to manage water security and could be applied more systematically to improve water security cost-effectively.

Following a risk-based approach to water security, a risk is considered *acceptable* if the likelihood of a given hazard is low and the impact of that hazard is low. In such cases, there is no pressure to reduce acceptable risks further, unless more cost-effective measures become available. In contrast, cost-effective measures are required to reduce *tolerable* risks to an acceptable level. Due to their very high probability and/or high damage potential, *intolerable* risks require urgent action to reduce them to an acceptable level. The acceptability and tolerability judgement process enables policy makers to prioritise risk management decisions when risks exceed acceptable levels (OECD/Swiss Re/Oliver Wyman, 2009).

Achieving water security means maintaining acceptable risk levels for four water risks – see risk terminology in the glossary of terms in Annex A:

- *Risk of shortage (including droughts):* Lack of sufficient water to meet demand (in both the short- and long-run) for beneficial uses by all water users (households, businesses and the environment).

- *Risk of inadequate quality:* Lack of water of suitable quality for a particular purpose or use.

- *Risk of excess (including floods):* Overflow of the normal confines of a water system (natural or built), or the destructive accumulation of water over areas that are not normally submerged.

- *Risk of undermining the resilience of freshwater systems:* Exceeding the coping capacity of the surface and groundwater bodies and their interactions (the "system"); possibly crossing tipping points, and causing irreversible damage to the system's hydraulic and biological functions.

All four risks must be assessed at the same time as they can impact on each other given the nature of water as a hydrologically interconnected resource. Indeed, these risks are interrelated; for instance, the risks of shortage, inadequate quality and excess may all increase the risk of undermining the resilience of freshwater systems. Managing all of these water risks is central to achieving the objective of water security.

From an efficiency perspective, the management of these water risks should focus on events with the most impact. There is generally an assumption that this means focusing on extreme ("tail") events with low probability and high impact, such as extreme floods.[1] But the long-term catastrophic consequences of "normal" (low immediate impact) but recurrent or chronic threats to water security, such as competition or pollution, deserves much greater risk management attention. These concealed or dormant risks develop slowly and are thus often considered as "invisible", with their main impacts emerging only in the long term. Yet there are subtle signs that may signal risk triggers, such as slower recovery from small disturbances (known as "critical slowing down").[2]

To date, water risk management has largely focused on protecting critical assets from disasters, on emergency preparedness and short-term crisis management, and much less on long-term water security. Furthermore, risk assessment and management have been applied piecemeal to certain aspects of water management (e.g. drinking water standards, flood control) but have not covered water resource management holistically, from a risk perspective. Water resources are still not managed for risk, but for certainty (e.g. supply security, access security).

Yet, water management, at its core, is about *reducing or avoiding water risks* and about the distribution of residual water risks – who *bears the risk*. But water management decisions are often driven by imperatives such as economic constraints and opportunities, and the costs and benefits and distributional impacts of risk management are seldom expressly considered. Responses to water risks may *transfer risks* to others or defer them into the future. They may also *increase other water risks* (e.g. reducing the risk of water shortage may increase the risk of undermining the resilience of freshwater systems). Current policies often fail to recognise these unintended effects ("externalities") and to address these trade-offs between water risks (risk-risk trade-offs).

Most countries face seasonal or local water shortage problems and several have extensive arid or semi-arid regions where water is a constraint to economic development. Multiple and scattered ("diffuse") sources of water pollution are challenges in many countries and a multiplying number of water contaminants threaten freshwater quality. The population affected by flood is increasing worldwide and with it the value of assets at risk.

This report *Water Security: Managing Risks and Trade-offs*: i) calls for action to manage the risks to society and the environment of water shortage, excess and pollution and of fragile freshwater systems; and ii) looks for solutions to maximise expected social welfare of trade-offs to maintain an acceptable level of such water risks to all. The report focuses on OECD countries but also refers to countries outside the OECD area.

The water outlook

The outlook is not optimistic (see Annex B for more details). A 55% increase in world water use is projected between 2000 and 2050. By 2050, the *OECD Environmental Outlook* projects that more than 40% of the world population will live in river basins under severe water stress (i.e. in river basins where withdrawals exceed 40% of available resources). This means an additional 1 billion people compared with today. The projected degradation of water quality adds to uncertainty about future water availability. By 2050, flood risks are projected to affect nearly 20% of the world's population.

As the world population rises to an expected 9 billion by 2050, water risks will be exacerbated. The process of urbanisation will increase, along with the demand for food and energy, and the pressures on the environment. Water risks will also be exacerbated by the immeasurable effects of climate change, which will increase uncertainty.

OECD and non-OECD countries face different water security challenges. Even though water use and nutrient pollution are increasing at much faster pace outside the OECD area, particularly in the BRIICS, diffuse sources of pollution, seasonal or local water shortage and floods remain an issue in most OECD countries, as is financing to replace ageing infrastructure and meet increasingly stringent environmental and health standards.

Managing water demand to balance with the available supplies in a way which promotes sustainable development is a formidable global challenge. The problem will be greatest in non-OECD countries that, as a group, are expected to have much larger rates of population growth. This is particularly so in large developing countries, such as India, where the rate of increase in incomes is also expected to exceed the OECD average. By contrast, a move towards water pricing based on supply costs in urban and industrial sectors in the OECD area, together with water recycling investments and improvements in water use efficiency in agriculture, have resulted in decoupling water demand from GDP (Annex B).

Unsurprisingly, the water outlook differs significantly between OECD and non-OECD countries. Water demand is actually projected to decrease in the OECD area (from 1 000 km^3 in 2000 to 900 km^3 in 2050). The projected decrease in water demand is driven by efficiency gains and a structural shift towards service sectors that are less water intensive. It is doubtful, though, that this will be enough to address the serious regional water supply-demand issues that already exist in parts of Australia, Israel, Mexico, Spain and the United States.

In contrast, water demand is projected to increase significantly in the BRIICS (from 1 900 km^3 in 2000 to 3 200 km^3 in 2050) and to a lesser extent in the rest of the world (from 700 km^3 in 2000 to 1 300 km^3 in 2050). Most of the population in river basins expected to be under severe water stress live in the BRIICS.

Similarly, the projected increase in nutrient pollution from point and diffuses sources is more significant outside the OECD area.

Moreover, there is a massive gap in major water infrastructure between OECD and non-OECD countries, both in terms of water services, water storage capacity per capita and share of hydropower potential developed. Compounding the infrastructure gap, there is a "bad hydrology" problem. There is a strong correlation between rainfall variability and GDP (Brown and Lall, 2006). Rainfall in most rich countries is moderate and predictable, whereas many poor countries suffer more frequently from droughts and floods, face higher levels of inter-annual uncertainty and confront ever greater variability as the world climate changes. In that context, water excess and shortage risks are seen as a profound cause of underdevelopment. However, as will be explained below in the context of water stress, these global indicators mask local disparities and water security concerns.

But there are still significant water security concerns in the OECD area. For example, pollution loads from diffuse agricultural and urban sources are continuing challenges in many OECD countries. Most OECD countries will probably continue to face seasonal or local water security problems and several countries have extensive arid or semi-arid regions where water could continue to be a constraint to economic development. In addition, climate "weirdness" could result in OECD countries experiencing more irregular precipitation patterns, which could have important economic impact given the strong correlation between rainfall variability and GDP. Moreover, OECD countries will need to mobilise significant financial resources in the next few decades to replace ageing infrastructure and to meet increasingly stringent environmental and health standards. As seen previously in the context of water supply, water demand, water quantity and water quality, water security is an issue in both OECD and non-OECD countries.

Applying a risk-based approach to water security: A conceptual framework

A risk-based approach to water security can be informed by a conceptual framework laying out the three steps involved in risk assessment and management, namely "know", "target" and "manage" the water risks (Box 1.1). The utility of this conceptual framework is in providing a comprehensive view of the three steps and their traits, along with a depiction of information and decision making flows and interaction with key stakeholder groups.

The framework combines the typical elements of technical risk assessment with important contextual factors, such as risk perceptions and risk evaluation. These contextual factors influence the demand for risk management and the willingness to pay

Box 1.1. **A risk-based framework for water security**

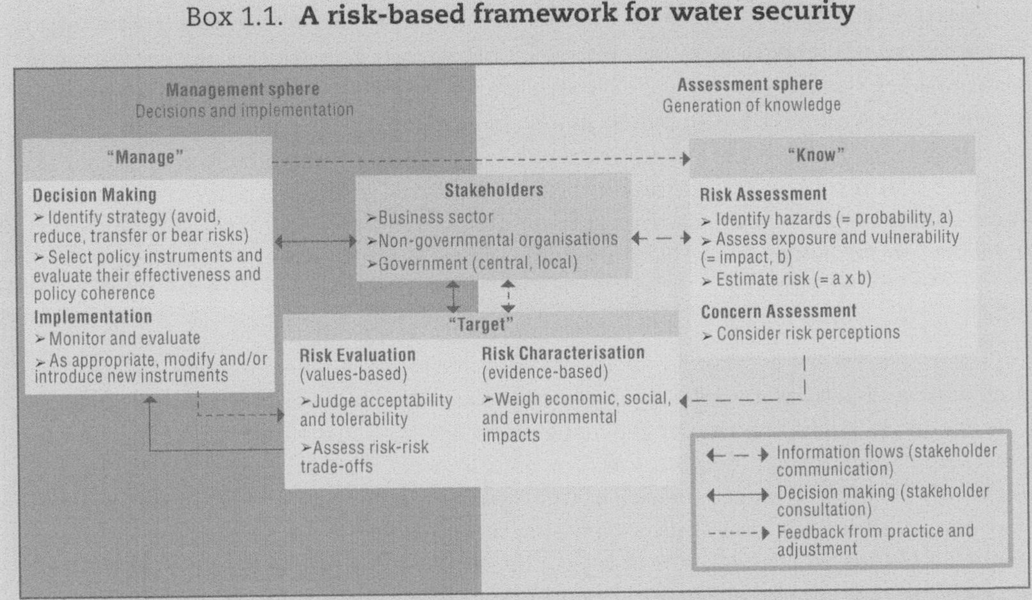

Source: Adapted from Renn and Graham (2006).

Know the water risks

Improving knowledge and reducing information asymmetry are the basis for making effective and informed risk management decisions. Yet, there is a striking lack of information on water risks and their scale (information gap), compounded as water resource management enters an era of uncertainty, greater variability and increasing risks as a result of climate change, population pressures, increasing demand to meet environmental needs and other risk drivers.

Although good science and technical expertise are needed, the understanding of risk perceptions via a *concern assessment* is a fundamental (and often overlooked) step in the risk appraisal process. It is a key element in seeking to assign roles and responsibilities for managing water risks. Indeed, individuals or businesses' perception of risk has an important influence on their decisions affecting their vulnerability to the risk and risk management strategies.

Appraising water risks means identifying areas subject to high-severity events, including "tail events" (i.e. low probability/high impact risks), but also "slow-developing catastrophic risk" areas, which are subject to low but cumulative impacts (e.g. gradual depletion of water resources; accumulation of pollutants in sediments).

Set acceptable levels (targets) for water risks

Achieving water security requires maintaining acceptable levels of risk – in terms of water shortage, excess, pollution, and freshwater system resilience – for society and the environment, today and in the future, through the effective and efficient application of water and water-related policies.

A water risk is considered *acceptable* if the likelihood of exceeding a given risk threshold (e.g. river flow, health standard, flood magnitude, tipping point of a freshwater system) is low and the impact of exceeding that threshold is low. In such cases, there is no pressure to reduce acceptable risks further, unless more cost effective measures become available. In contrast, cost effective measures are required to reduce *tolerable* risks to an acceptable level. Due to their very high probability and/or high damage potential, urgent action is needed to reduce *intolerable* risks to an acceptable level.

Box 1.1. **A risk-based framework for water security** (*cont.*)

The demand for risk reduction is influenced by factors beyond the risk profile itself determined by the severity of negative impacts and their likelihood. The (bold) claim that "acceptable" levels of risks should be determined only by scientific information about hydrology to the exclusion of any other criteria is a weak one. Although discussions of risk in water planning have traditionally been dominated by uncertainty in hydrology (increasingly so with concerns over climate change), due attention must be given to economic, social, cultural and environmental factors, which can be more important than hydrological uncertainties.

The acceptable level of water risk for society and the environment should depend upon the balance between economic, social and environmental consequences *and* cost of amelioration. Indeed, water security can be improved – but only at a *cost*. This cost may be in economic (e.g. building new or replacing old water infrastructure), social (e.g. closing water allocations to cap demand) and/or environmental terms (e.g. deterioration of freshwater systems to reduce the risk of water shortage). Depending on the existing level of water security, incremental improvements may, in some cases, be disproportionally costly. By identifying the level of acceptability of risks, a risk-based approach fosters targeted and proportional policy responses and thus cost effectiveness. As a result, targets for water risk vary between uses of water.

A tool to inform *trade-offs between policy objectives* is to document all the uses and associated values. Water use can be part of the market economy, making valuation of use relatively straight forward, or it can be non-market uses which are more challenging to value. Setting targets for water risks should thus be transparent about values and their trade-offs and consider equity between users.

Manage the water risks

Decision making about the appropriate response to water risks and the implementation of actions build on all the previous steps of the risk management process. The *risk management* strategy may be to avoid, to reduce, to transfer or to bear the risk. This can be done by altering risk drivers, limiting exposure or making populations, ecosystems and activities less vulnerable to potential harm. In cases where a policy response is considered appropriate, policy options should be assessed from an economic, environmental and equity perspective, to ensure that risk reduction is proportional, pursued at least cost and at least distributional impacts.

A risk-based approach allows for assigning risks to the actors able to manage them most efficiently in social welfare terms. For example, flood risks may be addressed more cost-efficiently through flood insurance or compensating farmers converting their land into flood plain instead of government investing in dams. Overall, the rational expectation would be that governments would only take direct action if, for instance, risks were collectively consumed on a large scale, when the potential for risk transfer was significant, where individuals had highly constrained opt-out options and where individuals or communities were deterred from making private safety provisions because they could not exclude free riders.

Maintaining acceptable levels of risk ultimately means addressing *trade-offs between policy instruments*. This requires a coherent approach between water policies and other (sectoral, environmental) policies.

for a given amount of risk reduction. The framework explicitly recognises "know", "target" and "manage" as a process driven by both evidence-based and value-based judgements.

Know the risks

This step first entails framing the risks by identifying the main drivers impacting on the hazards, exposure and vulnerability and projecting their long-term trends. **Drivers of water risks** include socio-economic trends, natural phenomena and inadequate water and water-related policies. Demographic and socio-economic trends, such as population growth and economic activity may strain water resources via increased abstractions and pollution. Urbanisation and decisions about land-use may increase exposure to water risks, including the risk of excess water, and to hazards such as natural disasters (e.g. earthquakes). Natural climate variability and climate change generate and exacerbate weather-related hazards. Social and cultural factors are also important risk drivers as they influence risk perceptions and may exacerbate man-made disasters and crises (e.g. terrorism, conflicts).

But water policy itself is what drives water risks the most (Grafton, et al., 2012). For example, it may lead to a lack of adequate water infrastructure and technology, due to neglect, insufficient financing and/or poor management and maintenance. Water risks are also the result of spillover effects. By creating incentives towards meeting their own security objectives, sectoral (e.g. agricultural, energy) and environmental (e.g. climate, biodiversity) policies have significant spillover to water security. For example, by distorting production and trade of agricultural commodities, agricultural policy distorts the domestic demand for water.

Building an adequate information base to inform decisions about water risks then requires appraising water risks through bringing together two components – a scientific risk assessment as well as an understanding of risk perceptions by stakeholders. The aim of the **risk assessment** process is to produce a best estimate of the physical harm a water risk may cause as well as identifying the exposure and vulnerability of populations, ecosystems and activities. The outcome of a formal risk assessment is an estimation of the risk in terms of a probability distribution of the modelled consequences.

A formal risk appraisal process for water risks can be data-intensive and costly in terms of time and resources. Often, significant scientific capacity is needed. In cases where significant populations, ecosystems and activities are at risk (e.g. densely populated urban areas in flood plains) and the cost of risk reduction is significant (e.g. structural flood protection), a formal and comprehensive risk appraisal is justified. In other cases, a less formal, but still informative, qualitative assessment or rapid risk assessment may be sufficient. The depth and extent of the appraisal undertaken should be proportional to the magnitude of the risk.

Although good physical science and technical expertise are a prerequisite for sound risk management, they alone cannot be the main basis for decision making (Rees, 2002). The understanding of risk perceptions via a **concern assessment** is a fundamental step in the risk appraisal process. It is a key element in seeking to assign clear roles and responsibilities for managing risks. Indeed, individuals or businesses' perception of risk has an important influence on their decisions affecting their vulnerability to the risk and risk management strategies. Concern assessment also helps solving the issue of "contested values"[3] that are often at the heart of conflicts over water (e.g. determining sustainable levels of use and/or pollution). The conventional risk assessment process, which typically places scientific assessment at the starting point for analysis, could even be inverted to place the human context, knowledge, needs and preferences as the first stage of appraisal. Such "inverted' risk appraisal models would appear to have a useful role to play to ensure that risk management becomes more demand responsive and more inclusive in terms of risk management options.

In appraising water risks, there will be many cases where the scientific assessment will be limited by sparse data, knowledge gaps and other sources of **uncertainty**. This is particularly acute when taking into account the impacts of climate change, for which confidence levels are often low for key climate parameters (OECD, 2013). A common distinction between risk and uncertainty derives from Knight's (1921) observation that risk is uncertainty that can be reliably measured. Thus, risk describes the likelihood and consequence of an uncertain event of which the probability of occurrence can be reliably estimated. Uncertainty describes situations where the probability of occurrence is not known and perhaps cannot be known. The difference between risk and uncertainty can be understood as a spectrum, where uncertainty is an expression of the degree to which a value or relationship is unknown.

A key step to dealing with uncertainty in risk assessment is to identify the sources of uncertainty and to be explicit about the degree of confidence experts have in the scientific knowledge base. Uncertainty can be characterised quantitatively, for example, by a range of values calculated by various models assessed with various confidence intervals; or qualitatively, by reflecting expert judgement.

The **role of the government** is first and foremost to facilitate the provision of information to improve knowledge and reduce information asymmetry as the basis for making effective and informed risk management decisions. Indeed, there is a striking lack of information on water risks. The knowledge, science and monitoring of hydrology, environmental and water resource management linkages is less well developed than have been the advances in water policies in many countries (OECD 2010). This disconnect means that decision makers are poorly informed and that policies are inadequately implemented and evaluated. These gaps in knowledge, science and monitoring are compounded as water resource management enters an era of greater uncertainty, variability and higher risks as a result of climate change, population pressures, increasing demand to meet environmental needs and other risk drivers.

Information failures are a main source of disparities in the distribution of water risks due to imperfect knowledge and information asymmetry (those exposed or vulnerable to risks lack the knowledge to make informed choices about their own welfare). Because of unequal distribution of information, information asymmetry creates risk transfer externalities.[4]

Moreover, there is critical lack of data and information on the economic aspects of water management. Furthermore, the provision of water security to one community or group of users can create the perception of relative disadvantage to other communities. The understanding of such risk perceptions is too often overlooked.

As new information on water risks develops, it may help resolve ambiguity about the way risks are shared between users and the government. For example, the risk of reduction in water availability in the Murray-Darling Basin is to be borne by users if it is due to new knowledge about the hydrological capacity of the system, and by the public if it arises from changes in public policy, such as changes in environmental policy (Quiggin, 2011). In the latter case, water users will receive compensation for such reductions in available water.

Target the risks

Based on the results of the risk appraisal, the process to determine the appropriate response begins by determining the "acceptability" of the risk. This relies on both evidence- and values-based judgements. Economic analysis has an important role to play for the evidence-based judgement, along with analysis from natural and social sciences. The overarching purpose of the **risk characterisation** process is to produce the best possible estimate of the broader economic, social and environmental implications of the risk.

The **risk evaluation** consists of making the distinction between acceptable, tolerable and intolerable risks (Figure 1.1). This is one of the most challenging and controversial tasks in the risk management process (Klinke and Renn, 2012). Indeed water security touches upon the issue of allocating water risks between residential, agricultural, industrial and environmental uses, a significant political economy question, as each will define *essential* or *adequate* in different ways.[5] For instance, much of the current policy debate in Australia's Murray-Darling Basin is about reallocating water from irrigation to sustaining the ecosystems. The reallocation of water among users can be seen, in effect, as a reallocation of water risks. In this example, the shift in allocation increases the risk of shortage to irrigators in an effort to decrease the risk to the resilience of freshwater systems.

Figure 1.1. **Acceptable, tolerable and intolerable risks**

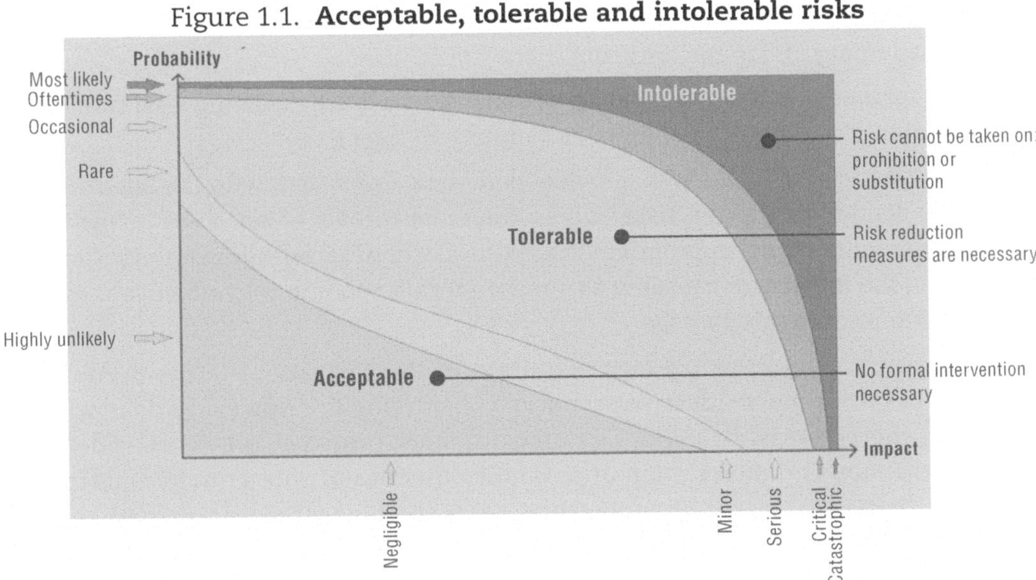

Source: Klinke and Renn (2012).

Societies vary in the ways in which they select which problems are identified as risks, and which risks require attention and response, and what is an acceptable level of risk. Different actors within societies define risks and levels of acceptable risk in different ways. Too much or too little water can be considered a risk, depending on the level of water required.

The risk evaluation process also characterises potential **risk-risk trade-offs**. Indeed, efforts to reduce water risks for a given population, ecosystem or activity may (inadvertently or not) increase other water risks. For example, reducing the risk of water shortage through increased diversions can increase the risk of undermining the resilience of freshwater systems. Risk-risk trade-offs depend in fact on how water risks are managed. For example, should the risk of water shortage be addressed through improving efficiency of water use, this would have no effect on diversions and hence, on the risk to freshwater systems resilience. In many cases, risk-risk trade-offs will not involve choices between only two water risks, but several.

Weighing risk-risk trade-offs thus helps identify strategies that minimise the negative externalities of risk management. Risk-risk trade-off analysis helps policy makers evaluate the impact on water risks of policy intervention (or lack thereof), weigh the comparative importance of managing interrelated risks when difficult choices are required, and analyse the possibility of overall risk reduction (Graham and Wiener, 1995).

Addressing the trade-off between water risks can reduce inefficiencies and inequities. Weighing risk-risk trade-offs requires both scientific and value judgements, as illustrated in Box 1.1.[6] Criteria for making judgements about risk-risk trade-offs will likely include the magnitude of the risk, in terms of the severity of negative impacts and the probability of those impacts; the size of the populations, ecosystems and activities affected by each risk (in the case that different populations, ecosystems and activities are impacted) as well as distributional aspects related to the characteristics of the affected populations.

Government has a responsibility to facilitate stakeholders' agreement on the acceptability of water risk(s) in exposed and vulnerable areas (areas at risk). Indeed, the level of acceptable risk is a key cost driver for water security. A valuable (and unique) feature of the risk-based approach is to make this explicit and consider it in light of the costs imposed (today and in the long run).

Setting targets for water risks should be consistent with the existing legal framework (e.g. water quality standards, maximum and minimum river flow).

Governments should not aim to provide "zero" risk. Consistent with the scientific and technical understanding of the risks, where there are threats of serious damage to the water environment, it is not appropriate to use the lack of full scientific certainty about the magnitude of the impacts or causality as a reason for postponing cost-effective measures to prevent or minimise this damage.

The precautionary principle (see definition in the glossary of terms in Annex A) essentially assumes the worst-case scenario, considering the downside risks without considering the potential benefits. The precautionary principle is best considered in relation to the standard prescription of normative theories of choice under uncertainty, namely, to choose the course of action that yields the highest expected (net) benefits (Quiggin, 2005). The burden of proof of safety is on those who create risks (in contrast with the distribution of costs and benefits, which places the burden of proof on regulators to identify the risks).

More attention should be paid to the systematic assessment of the costs and benefits of reducing risks across the water use sectors and to the consequent evaluation of various risk trade-off options (Rees, 2002). Information about the water risks should include the methods and costs of reducing exposure and vulnerability or of adopting loss sharing schemes.

The demand for risk reduction (or, in other words, the acceptable level of water security) is influenced by factors beyond the risk profile itself determined by the severity of negative impacts and their likelihood. The (bold) claim that "acceptable" levels of risks should be determined only by scientific information about hydrology to the exclusion of any other criteria is a weak one. Although discussions of risk in water planning have traditionally been dominated by uncertainty in hydrology (increasingly so with concerns over climate change), due attention must be given to economic, social, cultural and environmental factors, which can be more important than hydrological uncertainties.

A good example of economic factor influencing flood risk acceptability is given by agriculture along banks of the Nile in Ancient Egypt. "Moderate inundation' was defined as the key element of agricultural productivity (and related tax revenues). A lighter inundation than normal would cause famine, and too much flood water would be equally disastrous, washing away much of the infrastructure built on the flood plain (Figure 1.2).

Figure 1.2. **Interpretation of readings from the Nilometer**

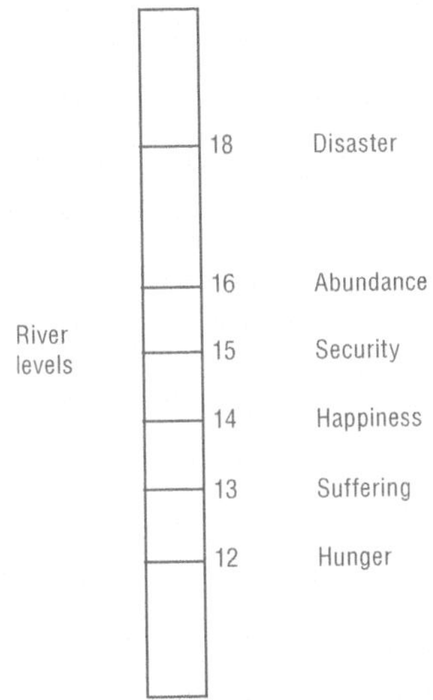

Source: waterhistory.org,. *www.waterhistory.org/histories/cairo/cairo.pdf.*

An example of social (health) factor influencing quality risk acceptability is given by the level of chlorine in drinking water. Because the key objective of public policy is to improve health and because disinfection by-products that result when chlorine interacts with organic matter may contribute to increased cancer, it is legitimate to ask whether levels of chlorine used in water treatment to reduce the likely of water-borne diseases should occur at the expense of potentially increasing the risk of cancer.

The reduction of (or failure to reduce) a water risk may occur at the expense of (generate) another water risk. Such trade-offs between water risks provide a good example of environmental factors influencing the acceptable level of risk reduction.

The demand for risk reduction is also influenced by cultural factors. For instance, a community may increase its demand for flood protection because neighbouring communities have benefited from such protection, instead of as a result of increasing frequency or severity of flood events.

Regulating the public provision of water security often reinforces the demand for it. The availability of water or protection from floods can provide incentives to increase exposure and vulnerability to water risks. For instance, flood protection may provide incentives for further development of flood plains. Over time, the risk of flooding may increase, in some cases, significantly. This increased risk shifts the cost-benefit assessment of flood protection significantly, as does the continued development in the floodplain.

Manage the risks

Decision making about the appropriate response to water risks and the implementation of actions build on all the previous steps of the risk management process. The **risk management strategy** may be to avoid, to reduce, to transfer or to bear the risk. This can be done by altering risk drivers, limiting exposure or making populations, ecosystems and activities less vulnerable to potential harm. In cases where a policy response is considered appropriate, policy options should be assessed from an economic, environmental and equity perspective, to ensure that risk reduction is proportional, pursued at least cost and minimise the distributional impacts.

The periodic monitoring and evaluation (M&E) of risk management strategies and tools provides for necessary adjustments and/or introducing new risk management instruments. Besides, the M&E results should feed back into the risk management process, as part of an iterative process, and may lead to revisiting the risk appraisal and/or reconsidering the acceptable levels of water risks.

Ideally, risks should be assigned to the actors able to manage them most efficiently in social welfare terms. For example, flood risks may be addressed more cost-efficiently through flood insurance or compensating farmers converting their land into flood plain instead of government investing in dams.

Yet, the impact of water management decisions on the **distribution of water risks** is seldom expressly considered in policy decisions. Public and private decisions that significantly influence the drivers of water risks as well as the exposure and vulnerability of populations and ecosystems to those risks are often driven by other imperatives, such as economic constraints and opportunities. As a result, the distribution of water risks is often characterised by inefficiency and inequity (Rees, 2002).

The treatment of water risks is often starkly uneven. On the one hand, the cost of reducing or avoiding risks can be unacceptably high (disproportionate to the risks avoided) (e.g. oversised urban wastewater treatment plants that were built in the new German *Länder* after 1990). On the other hand, potential threats can be overlooked entirely (reflecting difficulties to set acceptable risk levels) (e.g. there are still many priority substances that have yet to be regulated under water legislation).

Responses to water risks may transfer risks to others or defer them into the future. For example, flood protection may transfer flood risk from one community to a neighbouring one. A strategy to manage the risk of water shortage by unsustainably mining groundwater may simply transfer the risk from current to future users. We often undertake actions that inadvertently increase the risk of a disaster or magnify disaster losses at a faraway location or at some point in the future. We often fail to recognise these actions, and when we do, traditional approaches only poorly control these negative externalities (Berger, 2008).

The **role of government** in managing risk can be guided by several economic characteristics specific to water risk (in part, following Rees, 2002).

First, there may be a government's role when many people are affected by a water risk at the same time ("joint consumption of risk"). For example, all flood plain dwellers are exposed to the same potential hazard (though all may not be equally vulnerable).

It is also important to look at the geographical scale of such "joint risk". The rational expectation is that national governments would adopt the subsidiarity principle for spatially confined issues. If, for example, a pollution risk was confined to one locality and provided information asymmetries have been addressed, it would be possible for the government to enable the use of market-based instruments or dialogue between the polluters and those bearing the risk, rather than employing national coercive quality standards on the discharges of all polluters in the country.

Another (distinct) factor is the possibility and ease of opting out of some or all of the risk. If vulnerable people cannot avoid the risk *at all* (too costly, not physically possible) this could also be an argument for the government having a role. It has to do with the extent to which people can affect their risk consumption by choice (e.g. by altering their land use or building dwellings on platforms or by purchasing insurance). Where opt out is possible and easy the government *should not* manage the risk – because it could create moral hazard (i.e. increase the incentives that individuals have to take risks).

Enabling individuals to make their own decisions about risk mitigation measures is only likely to be effective if it is possible to physically exclude free riders who have not contributed to the safety provision. However, this clearly has ethical and equity implications, particularly when ability to pay is a factor behind the failure to contribute (e.g. provide no assistance to those without insurance). In practice where physical excludability is possible (e.g. deny access to a clean water source), societies will need to make judgements about whether the poor should be protected.

The degree to which risks can be transferred also matters (see definition of "risk transfer' in the glossary of terms in Annex A). This could occur within the water sector or produce risks in other sectors (e.g. reduction of risks from water pollution increasing risks from air, ocean or land pollution). In either of these cases, the greater risk transfer, the less likely governments would allow private choices to operate in an unregulated way.

Where the risk for some people magnifies the risk for others (e.g. risk spread in the case of water-borne disease), risk is in this case a public "bad" which the government needs to mitigate or regulate directly.

Overall, the rational expectation would be that governments would only take direct action if, for instance, risks were collectively consumed on a large scale, when the potential for risk transfer was significant, where individuals had highly constrained opt-out options and where individuals or communities were deterred from making private safety provisions because they could not exclude free riders.

As discussed in detail in Chapter 2, a risk-based approach to water security offers a new way to approach water policy making. Enhancing water security should address first and foremost how resource and pollution-related risks should be managed in light of the costs they impose and the expected benefits from improved management. A risk-based approach has the potential to facilitate:

- A *holistic approach to water security*. First, water security means addressing all water risks at the same time because they are interrelated. Second, setting acceptable levels of water risks means addressing trade-offs between water security and other (sectoral, environmental) policy *objectives*. Third, maintaining acceptable levels of risk means addressing trade-offs between policy *instruments*.

- *The assessment of policy priorities*. Indeed there is no need to address water security everywhere, in particular where the likelihood and the impact of water risks are low. By identifying areas at risk ("weak spots"), a risk-based approach can help to prioritise policy action, focusing on where to get more value for money. It can also ensure that risk management is proportional to the risk faced. Emphasising the proportionality of action to address risk can help to avoid taking action where the marginal cost of risk reduction exceeds the marginal expected benefits.

- *Preventive action.* A risk-based approach is a move from reactive policies (responding to pressures on water) to proactive policies (identifying where an impact might occur) based on the drivers of exposure and vulnerability to risk, which are not typically addressed in conventional approaches to hazards.

- *Dealing with uncertainty.* Uncertainty is inherent in the notion of risk. A risk-based approach allows for addressing uncertainty in a systematic and explicit way.

- *More responsive decision making.* The risk-based approach allows flexibility and responsive decision making. The acceptable or tolerable levels of water risks are not (should not be) static and will change over time, reflecting changes in risk drivers as well as in risk perceptions and water valuation. Feedback from practice is an integral part of risk appraisal and acceptability judgement, as part of an iterative process.

- *Long-term vision.* At times, solving urgent water security concerns requires short-term solutions. However, water security is not only about addressing immediate concerns but foremost to reduce risks of water insecurity over the long-term. Water security should be seen as a long-term goal.

- *Fostering equity.* By explicitly considering the distribution of water risks, a risk-based approach helps to prevent particular stakeholders to impose their own risk preferences on others or to gain benefits from risk management at the expense of others.

- *Ecosystems protection.* A risk-based approach seeks to achieve acceptable levels of water risk for society *and* the environment. The acceptable levels of water risks are set based on environmental quality objectives (e.g. water quality standards, maximum and minimum river flow).

- *Enhancing resilience.* By explicitly considering the risk of undermining the resilience of freshwater systems, a risk-based approach aims to develop water management practices that enhance such resilience.

Implementing the risk-based approach

As discussed in detail in Chapter 3, implementing a risk-based approach to water security will require governments to use a mix of policy instruments. Market-based instruments can play an important role in this policy mix as they can fundamentally alter the incentives facing water users, provide explicit signals about the likelihood and potential cost of water risks, and provide mechanisms for offsetting risks. This can be considered in the context of water supply, demand, quantity and quality.

As water risks are interlinked and the use of market-based instruments can have wider environmental and social impacts, a focus on economic efficiency by itself is not sufficient to tackle water security problems. Environmental and social goals need also to be considered. A widely accepted framework to implement this integrated approach is through integrated water resource management (IWRM), which encourages a more flexible, adaptive approach to water security management, involving greater collaboration with stakeholders and increasing the chance of sustainable outcomes to water security problems in the long term.

As discussed in detail in Chapter 4, managing water risks should be the result of well-informed trade-offs between water security and other (sectoral, environmental) *policy objectives*. Setting acceptable levels of water risks among stakeholders is one of the most challenging and controversial tasks in the risk management process. Indeed, allocating water risks between residential, agricultural, industrial and environmental uses raises a significant political economy question. Taking a broader view on interconnected and sometimes conflicting policy objectives, such as tensions between food security (and the willingness to secure domestic production) and water productivity (and the allocation of water to activities which add more value), trade-off choices can be improved.

Managing water risks has also to do with managing trade-offs between *policy instruments*. This requires a coherent approach between water policies (as described above) and other (sectoral, environmental) policies. Enhancing overall efficiency in water risk management entails taking account of complex links with sectoral policies, such as agriculture and energy, and other environmental policies, such as climate and nature.

By creating incentives towards meeting their own objectives, sectoral and environmental policies may have significant spillover to water security. The links between water and other related security objectives – food, energy, climate, biodiversity – are not routinely addressed or fully understood. Yet uncoordinated policy aimed at security in one area may result in less security in another: less water security as the cost of greater energy security through biofuel production, for example.

Complexity arises from the need to consider not only the direct but also the *indirect impacts* of sectoral policies on water security. The same sectors (e.g. agriculture, energy) that impact on water also impact on other components of the environment (e.g. climate, nature). Moreover, within a sector, the objectives of environmental protection and improving water management sometimes conflict with each other (e.g. subsidies to fast-growing forest plantations aimed at carbon sequestration are sometimes at the detriment of old growth natural forests that better regulate water flows).

What are the costs and impacts of inaction?

The costs of policy inaction can be considerable, not least because water insecurity can have global impacts (see Annex C). This is particularly the case where water insecurity causes disruptions in globalised businesses' supply chains. Not only are water risks directly affecting users (e.g. through the depletion of water resources), they also can result in significant additional use costs (e.g. increased abstraction costs due to groundwater subsidence). Moreover, there can be costs associated with damages to non-use values, such as the life-support function of water.

Inaction can thus lead to significant costs to society and the environment. Some costs of inaction are already reflected in household, firm and public expenditure (e.g. expenditure on health or to secure access to clean water or flood protection). Some are not, including the costs associated with biodiversity loss, though their impacts (in terms of lost welfare) can be significant. Loss of biodiversity reduces social welfare if the loss to society as a whole outweighs the gain to society (resulting from its loss), including it medium and long-term effects.

There is concern that segments of the population face greater exposure to water risks because they are more vulnerable (e.g. children), more exposed (living in areas at risk) and have more limited access to water resources and services (e.g. poorer households). In particular, microbial water pollution mostly hurts children and groundwater shortage the rural poor.

There is also a concern that disparities in water risks increase income disparities. Because they invest less in water security and are often living in areas at water risk (e.g. areas of poor water quality), lower income groups are more exposed to water insecurity and potentially "pay" a higher share of the costs of policy inaction (e.g. health costs) than higher income groups. In addition, water insecurity can marginalise those who lack access to capital (e.g. to invest in well-deepening as a result of falling water tables).

Notes

1. A tail risk is at the tail end of the risk distribution, with the least probability of occurring.

2. When a system is close to a tipping point, it can take a long time to recover from even a very small disturbance.

3. As a society, people value water highly for a range of economic, environmental, social, and cultural benefits, which at times are in conflict with each other (Bark et al., 2011).

4. Information asymmetry also hinders risk insurance initiatives.

5. Water security can be interpreted in terms of minimum levels of water risks for ensuring service provision which can be said to be "essential" (i.e. basic needs vs. "luxury" use), where one person's luxury use may be another's basic use.

6. There are techniques (e.g. stated preference or choice modelling) which allow for the expected benefits of reducing one type of risk to be weighed against both the expected costs of reducing that risk and the relative deterioration in safety from other forms of risk.

References

Bark, R. et al. (2011), "Water Values", in *Water, Science and Solutions for Australia*, Chapter 2, CSIRO Publishing, Collingwood, Victoria, Australia.

Berger, A. et al. (2008), "Obstacles to Clear Thinking about Natural Disasters: Five Lessons for Policy", in *Risking House and Home: Disasters, Cities, Public Policy*, Berkeley Public Policy Press, Berkeley, California.

Brown, C. and U. Lall (2006), "Water and Economic Development: The Role of Variability and a Framework for Resilience," *Natural Resources Forum*, Vol. 30, No. 4, Blackwell.

Grafton, R.Q. et al. (2012), "Global Insights into Water Resources, Climate Change and Governance", *Nature Climate Change*, Vol. 3, March 2013.

Graham, J.D. and J.B. Wiener (1995), "Confronting Risk Trade-offs", in *Risk vs. Risk. Trade-offs in Protecting Health and the Environment*, pp. 1-41, Cambridge, Ma., Harvard University Press.

Klinke, A. and O. Renn (2012), "Adaptive and Integrative Governance on Risk and Uncertainty", *Journal of Risk Research*, Vol. 15, No. 3, March.

Knight, F.H. (1921), *Risk, Uncertainty, and Profit*, Hart, Schaffner and Marx, Houghton Mifflin Company, Boston.

OECD (2013), *Water and Climate Change Adaptation: Policies to Navigate Uncharted Waters*, OECD Studies on Water, OECD Publishing, *http://dx.doi.org/10.1787/9789264200449-en*.

OECD (2010), *Sustainable Management of Water Resources in Agriculture*, OECD Studies on Water, OECD Publishing, *http://dx.doi.org/10.1787/9789264083578-en*.

OECD/Swiss Re/Oliver Wyman (2009), *Innovation in Country Risk Management*, OECD Studies in Risk Management, OECD, Paris.

Quiggin, J. (2011), "Managing Risk in the Murray-Darling Basin", in *Basin Futures, Water reform in the Murray-Darling Basin*, edited by D. Connell and R.Q. Grafton, The Australian National University, Canberra.

Quiggin, J. (2005), "The Precautionary Principle in Environmental Policy and the Theory of Choice under Uncertainty", Murray-Darling Programme, *Working Paper*, M05#3, University of Queensland, Brisbane.

Rees, J.A. (2002), "Risk and Integrated Water Management", *TEC Background Papers*, No. 6, Global Water Partnership, Stockholm.

Renn, O. and Graham (2006), *Risk Governance: Towards an Integrative Approach*, International Risk Governance Council (IRGC) white paper.

Chapter 2

Applying a risk-based approach to water security

This chapter provides guidance on how to apply a risk-based approach to water security through a three-step process: know the risks, target the risks and manage the risks. The chapter also provides insights on ways to adapt water risk management to the level of risk. By way of illustration, country cases of water risk management in selected OECD countries are included.

Applying the "know", "target" and "manage" framework

This section provides guidance to governments to help them implement the risk-based approach. It is not a series of ready-made prescriptions but rather a flexible tool that governments can use as a "checklist".

Know the risks

Increasing pressures on water resources and risks to society from inadequate water management are coinciding with increased recognition of the multiple uses and values of water. There is recognition now, for example, that some benefits from water resources rely on water being left in its natural environment to maintain water quality, and maintain ecosystems in a good condition to support a range of ecosystem services, such as food production from fisheries or spiritual values associated with water.

A tool to inform risks and trade-offs between policy objectives is to **document all the uses and associated values**. Water resources are valued for a range of uses, some of which consume water (as a good) and others which do not (such as commercial fishing and recreation where freshwater provides a service). Water use can also be part of the market economy, making valuation of use relatively straight forward, or it can be non-market uses which are more challenging to value.

For non-market uses, the use of ecosystem service frameworks can help elucidate values and assign a market value for use in cost-benefit analysis. It is increasingly recognised that many of the benefits that accrue from water are through ecosystem services: the role of water in maintaining rivers and wetlands in good ecological condition. Healthy rivers and wetlands provide services to society (such as maintaining water quality) on which can be placed an economic value (such as the avoided costs of water treatment).

But not all values are easily amenable to this valuation approach. Even if they were, experience with the recently approved Murray-Darling Basin plan in Australia shows that it will not necessarily resolve conflicts because different sectors of the community have different sets of values, effectively discounting some values over others. Conflicts over water use are often conflicts of competing values for water for which there are no agreed mechanisms to compare and resolve the various values (Bark et al., 2011).

Another tool to inform risks and trade-offs between policy objectives is to **assess risk sharing arrangements among sectors**. For example, reductions in water availability (such as during drought) are not equally shared by water users (cities, agriculture, industry, environment). Some users take a greater share of the water shortage risk as a result of the way water is allocated and managed. In some respects these uneven risk sharing arrangements are appropriate ways to deal with variable supplies and the varying values of uses. For example, urban and industrial water uses typically take a low risk, then agriculture takes larger risks (with uncertain supplies of irrigation water in the driest years) and lastly the environment and users downstream often bear the greatest risk. This is because the costs of not maintaining reliable supplies of high quality water to cities and industry are much higher than to irrigation of annual crops, where farming systems that include seasonal crop choices have evolved to successfully adapt to unreliable but inexpensive supplies of water. In other respects, though, the variable sharing of risks results in inequitable sharing of increased risks.

Climate change makes assessing the sharing of water risks more complex. Climate change does not introduce new types of water risks but increases or decreases water risks and introduces a greater degree of uncertainty. In some regions, climate change increases the overall pressure on water resources by decreasing water availability and increasing demand for water. In other regions, climate change can increase the magnitudes and likelihood of floods and pollution.

Developing the knowledge base on water risks does not necessarily entail sophisticated risk assessment techniques, which can result in a lengthy and costly exercise. As a principle, the **sophistication of risk appraisal should match the level of water risk**. Where there are billions of critical assets at risk, one can expect a strong and robust risk appraisal (e.g. flood risk assessment in big cities). In contrast, where current levels of risk are low, a basic risk appraisal can be used. For example, in Western Australia, where current demand for groundwater is low, a basic appraisal of groundwater abstraction risks has been used (Box 2.1).

Box 2.1. **Appraising the risk of undermining resilience of a groundwater system, Western Australia**

In Western Australia, for areas where knowledge of groundwater is limited and current demand for the resource is low, a risk-based approach has been developed to set water allocation limits and licensing rules (Government of Western Australia, 2011). The appraisal includes the risk to environmental, cultural and social groundwater-dependent values, as well as the development risks of not abstracting water for consumptive use.

The risk appraisal process has two steps:

- identify and define the groundwater resource (including estimation of aquifer recharge);
- describe aquifer properties, environmental, cultural and social groundwater-dependent values and assess the risks to those properties/values from abstraction; describe the consumptive uses of water from the aquifer and assess the development risks of not abstracting water for consumptive use.

Aquifers support groundwater-dependent environmental, social and cultural values and they can potentially yield water for productive use. It is important to maintain the quantity and quality of water in the aquifer; that is, the integrity (resilience) of the aquifer, so that it has the ability to yield water now and in the future. Risks to aquifer integrity can arise if abstraction alters the aquifer's water quality (e.g. through saline water intrusion). Risks to aquifer integrity can also occur through subsidence, where the removal of water from the aquifer leads to its compaction.

Groundwater-dependent ecosystems include wetlands, terrestrial fauna and vegetation, river baseflow systems, cave and aquifer systems, estuarine and near-shore marine systems. The level of risk to groundwater-dependent ecosystems from abstraction depends on how much they rely on groundwater, their sensitivity to changes in the quantity and quality of groundwater and their significance as ecosystems.

The cultural values associated with groundwater can be very high, particularly in environments where there are long periods without rainfall. In some areas, Aboriginal peoples' lives and belief systems are intimately linked with groundwater. Groundwater-

---

Box 2.1. **Appraising the risk of undermining resilience of a groundwater system, Western Australia** (*cont.*)

dependent vegetation and groundwater-fed pools and springs tend to be particularly significant for Aboriginal people. Vegetation, pools and springs may also have social values, depending on the degree to which they are accessible and used. As with groundwater-dependent ecosystems, the level of risk to cultural and social values posed by abstraction depends on their links with groundwater, their sensitivity to changes in groundwater and their significance.

A basic assessment of potential future consumptive use considers:

- the amount of water likely to be required;
- the degree of supply security required;
- the availability of alternative water supplies or alternatives to using any water at all within the production process;
- the purpose of the groundwater use;
- the social and economic benefits of the productive use.

The level of risk to consumptive use will be high where the socio-economic benefits of abstraction are high and there are no alternatives to using groundwater.

The overall risk appraisal aims at determining how much recharge can potentially be allocated. For this, ratings (high, medium or low) are assigned to *in situ* risk and development risk (see Table below). Risk rating considers the likelihood of abstraction impacting on the value (or the value's sensitivity to abstraction) and the consequences of that impact, or how important that particular value is. For example, a groundwater-dependent ecosystem may be highly sensitive to groundwater changes but of low environmental value, in which case its risk rating would be low. The highest risk ratings for *in situ* and development risks (respectively) are used to produce initial ratings.

	Values	Likelihood/sensitivity	Consequence	Risk rating	Overall risk
In situ risk	Aquifer properties	How sensitive is aquifer integrity to abstraction?	If aquifer integrity was to be impacted, how significant would that be?	High, medium or low	**In situ risk** (highest risk rating)
	Groundwater-dependent ecosystems	How dependent are GDEs on groundwater? What is the likelihood that GDEs would be impacted if water was abstracted, i.e. how sensitive are they to abstraction?	How significant are the GDEs in terms of environmental value?	High, medium or low	
	Cultural and social	How dependent are the cultural and social values on groundwater? What is the likelihood that these values would be impacted if water was abstracted, i.e. how sensitive are they to abstraction?	How significant are the GDEs in terms of cultural and/or social value?	High, medium or low	
Development risk	Current and future water use	How important is the resource for meeting current and future development needs? Are there alternative water sources or alternative production approaches that mean groundwater is not required?	How significant is the current and future productive use/ development for the community?	High, medium or low	**Development risk** (highest risk rating)

Source: Government of Western Australia (2011).

Target the risks

Achieving water security requires maintaining an acceptable level of water risks for society and the environment, today and in the future. The **setting of water security targets** can be guided by several economic characteristics (in part, following OECD, 2008).

By setting targets in ways that reflect water risks, water security makes important contributions to social welfare (e.g. by protecting the natural basis of production, and by improving human health). However, achieving the targets can also entail significant economic costs. It is therefore important to carefully consider whether the additional benefits of improved water security, and the additional costs to society of achieving these improvements, balance reasonably well. This implies the need to assess, on a regular basis, the costs and benefits of objectives that are set for water security. When feasible, this assessment should include monetary valuation of the changes in water security in question.

The setting of targets for risks is a difficult, but necessary, art. Non-linearities in the nature of water problems themselves (e.g. the risk of irreversibility), as well as uncertainty about the linkages between water risks and the economic values placed by producers and consumers on potential changes in water security, will make very complex any systematic effort to compare the costs and benefits of proposed targets. The economic value of water security improvement (or impairment) can be difficult to capture, especially when that value cannot be derived from direct use of water resources. Nevertheless, a comparison between costs and benefits still needs to be done, as one important input to decision-making, even when the water security outcomes of target setting are uncertain

The marginal costs and benefits of proposed water security targets should also be assessed on both an *ex ante* and an *ex post* basis. Likewise, the costs of policy inaction should also regularly be assessed. In conducting these benefit-cost assessments, the focus should be on final environmental "outcomes" (e.g. expected or actual improvements of water risks) and on the impacts of these outcomes (e.g. in terms of changes in health conditions) – rather than on intermediate "outputs" (e.g. the sharing of water volumes among stakeholders).

Economic values should also – to the extent possible – be placed on water security outcomes that are measured in physical terms. The idea is to quantify how much the public-at-large value changes in water security. This quantification can facilitate analysis and policy-making in situations where some water impacts pull in opposite directions (e.g. risk-risk trade-offs); it can also make it possible to compare the (private) costs and (public) benefits of a given target for water risk. In addition, the analysis should include qualitative descriptions and discussions of outcomes that cannot be quantified or monetised.

The "total economic value" of water security improvement (or impairment) includes both "use" and "non-use" values of that improvement. "Use" value refers to the direct benefits of actually using water (e.g. water withdrawn for irrigation). It also includes planned and possible future benefits of using the water resource. "Non-use" values refer to water that people will not actually use themselves at any point, but may want to preserve for others, for future generations, or simply because they attach a value to their very existence.

Putting a monetary value on water security is challenging, not least because not all the associated benefits do have a market value; they are also often not "tangible" assets. It is particularly difficult in situations when irreversibility and possible disastrous outcomes enter the equation, and when the resilience of water systems is being – or is close to being – overrun. In practice, most targets for water risk will have to be set without full knowledge of the benefits and costs involved. This is not an argument against trying to make as good an assessment as one can, but it makes it particularly important to take into account those cost and benefit elements that cannot be expressed in monetary terms.

The setting of water security objectives should ideally be done simultaneously with the setting of objectives in *all* policy areas of relevance – agricultural policy, energy policy, etc. Although it is clearly impossible to fully achieve this goal, it can at least be promoted by favouring governance structures that emphasise co-ordinated decision-making (e.g. the *systematic* review of major policy decisions by inter-Ministerial working parties).

The distributional impacts of targets for water risk (e.g. across income groups, age groups, ethnic groups or regions) should also be considered. The most common way of including distributional effects is to describe them separately from the cost-benefit analysis, without explicitly weighing costs and benefits affecting different individuals. When thinking about these costs and benefits, consideration should be given to all significant upstream and downstream impacts when setting targets for water risks.

In general, agreement on targets for water risks will be more likely if there is a common understanding of the problem at hand, its causes, and its impacts (over both the short- and long-terms), underpinned by correct information ("know the risk"). Governments should also obtain relevant information from stakeholders for the establishment of targets for water risks. This promotes both transparency and accountability. The acceptance of a given instrument by the public-at-large is strongly related to the degree of awareness of the water risk the instrument seeks to address, thus the importance of undertaking concern assessment.

The **acceptable level of water risk** for society and the environment should depend upon the balance between economic, social and environmental consequences *and* cost of amelioration (Figure 2.1). The limit of cost effective or practical water management is an element to consider when evaluating the cost of amelioration

Figure 2.1. **Setting the acceptable level of risks**

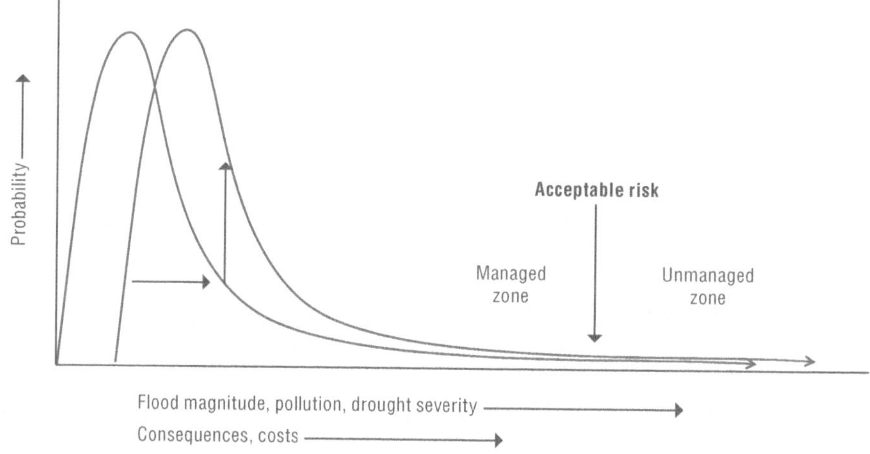

Notes: Event magnitude, (economic, social and environmental) consequences and costs (of amelioration) increase along the horizontal axis. The frequency (or cumulative effect) of events is shown in the vertical axis. Water management will be effective up to a certain size of event of low frequency (acceptable risk) beyond which the costs of management are deemed excessive. Larger events are dealt with as unmanaged emergencies.
Climate change may shift the distribution to the right (and change its shape). A small change in mean can result in a large increase in frequency of some events.
Source: Prosser (2012), inspired from Garrick and Hope (2012).

Targets for water risks should vary between uses of water. For example large dams might be built to survive a 1:1 000 year flood or probable maximum flood because the consequences of dam failure can be devastating to downstream populations. Residences and major roads might be built to avoid inundation from a 1:100 year flood, while minor roads and recreational facilities might only be secured from a 1:10 year flood. Surprisingly, New York City is protected to only a 1-in-100-year flood event, despite having a larger GDP than London, Shanghai, or Amsterdam, all of which are protected to a greater than 1-in-1 000-year flood (Amsterdam is protected from a 1-in-10 000-year floods) (Nicholls et al., 2008).

Similarly, for water supplies, urban potable water might be provided at a service level to meet demand in 95% of years and not cause any human sickness in 99% of years; whilst high security irrigation water for permanent horticulture might only meet demand in 90% of years and have lower water quality requirements such as salinity levels; and low security water supplies for annual crops and pasture might only meet demand in 50% of years and have a higher threshold of tolerable salinity.

Environmental water requirements can also take a similar form of percentage risk. Australian red gum floodplain forests on the Murray River, for example, require flooding for one month or more in 70% of years, while drier floodplain woodland ecosystems only require flooding for two months or more in 25% of years. Each use of water thus has a different level of acceptable risk (in this case a risk of shortage for the ecosystems that require periodic flooding).

Setting water security targets must deal with **uncertainty**. The more fundamental water security issue in terms of uncertainty is a non-stationary future (OECD, 2013). In particular, climate change has a range of complex effects on the global water cycle, including shifting the patterns of rainfall, increasing or decreasing water risks. In general, dry climates have the most variable river flows (Peel et al., 2004), posing higher levels of risk of water shortages in dry years. In many of the drier regions of the world a 10% reduction in rainfall could result in a 20-40% reduction in river flows (Chiew, 2006). Small decadal scale changes to rainfall in the last century have been observed to double the size of the 1:100 year flood.

Reducing such uncertainty through applying a risk-based approach (i.e. responding to continuous changes in knowledge as well as economic and socio-cultural conditions) can help improve the acceptable level of risks and with it generate higher economic returns.

For example, there are considerable uncertainties over the sustainable extraction limits for groundwater. Groundwater resources are renewed through recharge across the landscape but recharge rates cannot be measured at landscape scale. Groundwater then discharges through springs, lakes, rivers and directly to the sea which is also hard to measure, and these discharges are impacted by groundwater extraction. The most difficult aspect of sustainable groundwater extraction is the long lead times, up to decades, before over extraction is felt as lower pressure in wells or lower water tables. Recovery also takes decades. To avoid inadvertent impact on some users of groundwater, including environmental uses, use should be within the bounds of uncertainty over the resource.

As the pressure on the resource increases through greater extraction, investigations should be undertaken to reduce uncertainty and avoid the risk of over use. For example, rough water balance calculations for a groundwater resource might suggest that the mean annual recharge rate is 200 ± 100 GigaLitres/year [(GL)/y] and that groundwater dependent ecosystems need 30 ± 20 GL/y of water. At this level of uncertainty, neglecting other factors for ease of argument, only 50 GL/y of extraction could be safely allowed. If a groundwater model was built of the aquifer and recharge and discharge rates were calculated more accurately it might be determined that recharge is 190 ± 50 GL/y and ecosystems require 40 ± 10 GL/y. Then, extraction could increase to 90 GL/y. More detailed bore testing, dating of groundwater, and monitoring of water levels could reduce uncertainty further and allow extraction rates to increase further. In Australia these types of considerations are incorporated into guidelines for groundwater management.

As seen in the example above, the level of water use (e.g. groundwater extraction) is commensurate with the level of uncertainty or knowledge of the resource. The same applies to river water. Even in well gauged rivers there are significant uncertainties over water. These might include the sources and fates of pollutants in the basin; the effect of land use changes on flood magnitudes and propagation downstream, or the water requirements and tolerances to pollution of ecosystems.

As is the case of risk appraisal, setting targets for water risks does not obligatorily means sophisticated risk characterisation and evaluation techniques. Here also, **the level of risk characterisation and evaluation should match the level of water risk**. Following up on the example in Western Australia (see Box 2.1), where current demand for groundwater is low, a basic risk characterisation and evaluation has been used to set acceptable amounts of groundwater abstraction (allocation limits) and licensing rules (Box 2.2).

Box 2.2. **Setting targets for the risk of undermining resilience of a groundwater system, Western Australia**

The risk-based groundwater allocation planning process has two steps:

- Assess whether risks identified in the risk appraisal process can be managed through licensing rules.

- Set allocation limits (the amount of water available for consumptive use) and licensing rules.

First, there is a need to assess the capacity to manage the risks identified in the appraisal stage (see Box 2.1). For example, for some aquifers a high risk to groundwater-dependent ecosystems can be reduced to medium if the risk is managed through appropriate buffer zones. Licensing rules are defined at this stage and might include:

- Establishing buffer zones.

- Managing abstraction in relation to recharge events (e.g. reducing abstraction during droughts).

- Establishing triggers for additional management actions (e.g. groundwater levels).

Then the defined licensing rules are considered. If appropriate, the final ratings in Table 2.1 may be revised based on proposed mitigation measures. If risk mitigation strategies reduce the overall risk to *in situ* values, then the reduced risk value is used in the risk matrix instead.

A risk matrix is then used to convert the (final *in situ* and development) risks into a proportion of recharge (see Table below). To set the allocation limit, this proportion is applied to the estimated recharge volume defined in the initial step of the risk-based groundwater allocation planning process ("identify and define the groundwater resource"), which sets the volume ("target") that can be allocated for consumptive use based on an acceptable level of risk. The risk matrix allows consideration of the trade-offs between the two groups of risk.

> **Box 2.2. Setting targets for the risk of undermining resilience of a groundwater system, Western Australia** (*cont.*)
>
> | | Proportion of recharge | | |
> | --- | --- | --- | --- |
> | High *in situ* risk | 5% | 25% | 50% |
> | Medium *in situ* risk | 25% | 50% | 60% |
> | Low *in situ* risk | 50% | 60% | 70% |
> | | Low development risk | Medium development risk | High development risk |
>
> *Source:* Government of Western Australia (2011).
>
> This completes the target setting process with the outputs being:
>
> - An allocation limit.
> - A set of licensing rules to manage risks.
>
> The maximum allocation from this process is up to 70% of recharge. This allows for any uncertainty, given the limited information on aquifer properties such as recharge. Setting aside at least 30% of the estimated recharge protects the resource from potential over-allocation. It also protects aquifer integrity, including through reducing the risk of saltwater intrusion. The allocation limit can be revised as additional information becomes available.

Manage the risks

Once set, targets for water risks should be achieved at least possible economic cost (i.e. cost-effectiveness should be pursued). Increased water security and economic efficiency are necessary, but not sufficient, conditions for improved welfare. Another key dimension is the "social" dimension, including equity. Most people in society will feel their own welfare depends not only on their own individual exposure to water risks but also on the distribution of risks among citizens. When considering which particular instruments should be used to meet a given target for water risk, an assessment should be made of how much each instrument (or each "instrument mix") is likely to contribute to the goals of water security and economic efficiency (in part, following OECD, 2008).

Direct regulatory instruments (e.g. laws or regulations stipulating water quality standards or diversion limits, bans on certain products or practices, requirements for the application of "best available" techniques, obligations to obtain permits, etc.) represent a major proportion of all instruments currently being used for water policy in OECD countries, and they will continue to play a key role in the future. While the environmental effectiveness of direct regulatory approaches is often very good, the main challenge is to avoid undue inflexibilities in these regulations that might limit their environmental effectiveness and/or economic efficiency.

For government authorities, regulatory instruments are flexible – *inter alia* in the sense that they can be used to address a broad spectrum of water problems. They may also provide a relatively high degree of certainty about the environmental outcome – although this is not guaranteed. For example, although a particular water pollutant may be banned from the market, it is not always clear what product(s) will replace the banned item.

There are, however, a number of potential problems associated with regulatory instruments, from the perspective of economic efficiency. First, designing regulations places very significant information demands on public authorities – and some of the most relevant information is often only available from those who cause the water problem in the first place.

Second, even if the same standards are applied to all water users/polluters, these standards will not normally provide the same incentive at the margin for all water users/polluters to improve water security. From the point of view of the users/polluters, regulations can also be rather *inflexible* – because they sometimes impose a specific way of improving water security. If regulations are excessively inflexible, a given water security objective will not usually be reached at the lowest possible cost.

Third, whereas a water tax would provide a relatively high degree of certainty as regards the marginal compliance cost faced by users/polluters, a regulatory instrument does not provide similar certainty – even though careful *ex ante* assessments of expected impacts can also sometimes provide indications about the marginal compliance costs of regulatory instruments.

Fourth, both taxes and tradable permits give users/polluters a *continuing* incentive to improve water security through innovation, and therefore to develop new technologies. Regulatory instruments provide incentives to innovate (in order to reduce compliance costs) up to the point where users/polluters are in compliance, but they do not give any incentive to go *further* than this level. This disadvantage can to some extent be addressed by a gradual tightening of the regulations – but the cost of complying with the stricter requirements will again be unknown at the outset.

The cost-effectiveness of regulatory instruments broadly varies with the degree of *flexibility* they leave to users/polluters in responding to the regulatory requirements. A regulation which specifies that a certain technology has to be used leaves the user/polluter with very little choice, and can therefore trigger higher costs for improvement/abatement than necessary. Such an approach would also discourage innovation in potentially cost-saving alternative technologies – mainly because it would be unclear if the users/polluters would be allowed to use the new technologies. An abstraction/pollution standard which varies according to the type of product that is used (e.g. energy mix) could also eliminate incentives for users/polluters to switch to products that cause less water risk. Conversely, a regulation that focuses on the environmental outcome (e.g. diversion limit or ambient water quality standards) would provide more flexibility for users/polluters to find low-cost improvement/abatement options.

Therefore, new regulatory instruments should provide as much flexibility as possible for users/polluters to find low-cost improvement/abatement options. They should also not usually specify which technologies are to be used to reach a certain water security target. Regular consideration should also be given to whether existing regulations unnecessarily limit users/polluters' flexibility to apply existing cost-saving improvement/abatement options – or to develop new ones.

Direct regulations sometimes include more lenient provisions for small users/polluters than for large ones. For example, limits on water pollution from livestock facilities beyond a certain capacity are sometimes stricter than for livestock facilities with a lower capacity. Whereas special provisions for small sources can sometimes be appropriate (because small firms can have a lower capacity to finance expensive improvement/abatement or monitoring equipment), it is important to make sure that this kind of provision does not undermine the environmental integrity of the original regulation (e.g. by encouraging the establishment of a significant number of new water abstraction/pollution sources with a size just below the chosen limit, simply to avoid the application of the regulation). Small firms – when considered together – may also represent a large source of abstraction/pollution.

When new regulations are introduced, stricter provisions are often applied for new abstraction/pollution sources than for pre-existing ones. Although it can be economically efficient to give existing sources some time to adjust, special treatment of sometimes heavily-abstracting/polluting existing sources can prolong their economic life *beyond* what would have been the case in the absence of the regulation – because of the additional burden the regulation places on new sources. Hence, any preferences given to pre-existing abstraction/pollution sources should usually be time-limited.

Even with the most flexible forms of regulatory instruments, different users/polluters will normally face very different costs of reducing abstractions/pollution by an additional unit. Explicit consideration should therefore be given to the possibility that the regulatory approach could be partially replaced by market-based instruments (taxes or trading systems).

Water security objectives could be met in a more cost-effective manner by using market-based instruments, such as water taxes (e.g. abstraction taxes, pollution taxes). These **taxes** provide incentives for polluters and resource users to change their behaviour today. They also provide long-term incentives to innovate for a more water secure future tomorrow. Although water taxes are not strongly supported by the public in all contexts, there are various ways in which this support can be increased over time (e.g. through measures to limit negative impacts on the competitiveness of certain sectors and/or on income distribution).

Water taxes are increasingly being used in OECD countries, and there is good evidence of their environmental effectiveness in many cases. In the short term, these taxes can reduce the production and consumption of products whose manufacture and/or consumption impairs water security. In the longer term, they encourage the development of new production methods and new products that meet consumer demand, even while improving water security.

An advantage of taxes, compared to regulatory instruments, is that the former are often less demanding in terms of the information that public authorities need to have at their disposal, in order to be environmentally effective and economically efficient. On the other hand, the relative newness of water taxes means that, in practice, it will still be necessary to gather a significant amount of information about the expected impacts of these instruments, before an agreement on using taxes can be reached.

There is a high potential for greater use of water taxes, both in terms of new taxes that could be applied to environmentally harmful goods and services, and in terms of increases in existing tax rates, where these taxes already exist – in order to better reflect the environmental externalities of relevance. However, these taxes need to be well-designed and their potential impacts on international competitiveness and income distribution need to be fully addressed, if these benefits are to be realised. This is because taxes are relatively blunt instruments and can – if not used correctly – have negative consequences for the achievement of more specific policy objectives.

To enhance the environmental effectiveness and economic efficiency of water taxes, a key first consideration is the possibility of scaling back the exemptions and other special provisions already contained in existing water taxes, and to better align the tax-bases (i.e. the "object" that is being taxed) and the tax rates with the actual magnitude of the negative environmental impacts that need to be addressed.

The revenues from water taxes can be used to strengthen the budget balance; to finance increased spending; or to reduce other, distortionary, taxes. There are sometimes calls for water tax revenues to be "earmarked" to specific spending purposes – in some cases, to increase water spending ("water pays for water"). However, earmarking also raises a few problems. For example, it could actually violate the *Polluter-Pays Principle*, if the money is used to cover the additional cost faced by polluters for meeting water security requirements. Earmarking also fixes the use of the tax revenues, which may create an institutional obstacle for later re-evaluation and modification of tax and spending programmes. More generally, earmarking tax revenues for specific uses does not guarantee "value for money"; it also removes this revenue stream from other spending opportunities.

Water taxes are well suited to addressing problems such as reducing the *total* amount of a given type of abstraction/pollution (or the use of a given water consuming/polluting product) within the geographical area in which the tax is applied. However, taxes are less well-suited to addressing (on their own) site-specific problems (e.g. local "hot spots" of water shortage/pollution), and with situations where it matters *when, how* or *where* a certain water consuming/polluting product or practice is being used or implemented. In such cases, a water tax might need to be combined with additional instruments, such as standards on ambient environmental quality in different areas, regulations specifying conditions for the use of water consuming/polluting products, etc.

Water taxes *can* entail relatively low administrative costs. For example, taxes on hydroelectricity are levied on a limited number of hydropower plants, and are therefore relatively simple to administer and enforce. On the other hand, many taxes involve various special provisions that can significantly increase administrative costs. Such mechanisms are often introduced for non-environmental reasons (e.g. to address competitiveness or income distribution concerns). It will often prove to be more efficient and effective to promote fairness by using non-environmental policy instruments (e.g. the social security system or the income tax system), rather than by amending the conditions of the original water tax.

Closely linked to the use of water taxes are prices, fees, and charges for water services (e.g. water supply, waste water treatment) As is the case for taxes, the prices facing firms and households for these services should reflect the *full marginal social costs* of providing them.

It is also important to periodically review the actual performance of water taxes, in order to determine if further improvements could be made in their environmental effectiveness or economic efficiency.

Tradable permit systems provide similar flexibility as taxes do for polluters/resource users to choose the method by which they will achieve a given water security goal. By establishing "caps" or promoting direct investment in environmentally beneficial outcomes, they also emphasise the achievement of water security goals. Nevertheless, there are several issues that need to be considered when using this approach, in order to increase the environmental effectiveness and economic efficiency of permit trading (e.g. the choice between a "cap-and- trade" system and a "baseline-and-credit" system; the initial allocation of abstraction/pollution allowances; and ways of limiting the transaction costs associated with the trading system).

Like taxes, tradable permits provide a flexible, market-based, approach to the achievement of water security objectives. This flexibility helps to reduce the cost of abatement (both short- and long-term). On the other hand, unlike taxes, the water security objective is explicitly reflected in the number of abstraction/pollution permits that are issued, which means that this water security objective should actually be achieved. In fact, this is a key characteristic of tradable permits systems – they are quantity-based (not price-based) measures, which means that they focus mainly on the water security outcome, rather than on the economic cost of achieving that outcome.

Tradable permit systems introduce a quantitative limit in the form of either: i) a maximum ceiling (in the case of cap-and-trade schemes); or ii) a minimum performance commitment (in the case of baseline-and-credit schemes). These limits can also be expressed either in absolute terms or in relative terms, and the permits can be denominated either in terms of "bads" (e.g. pollution emissions) or of "goods" (e.g. access to water resources). When cap-and-trade systems are used, there is a high degree of certainty about the environmental effectiveness of the instrument – because the water security outcome is explicitly embedded in the cap that is chosen.

The total cap (in a cap-and-trade system) is of vital importance for the water security outcome of the scheme. These water security caps should be set at levels that are consistent with long-term water security objectives. As for any other form of environmental policy, these caps should seek to strike a balance between the long-term marginal costs and the long-term marginal benefits of the trading programme. In order to give firms and households time to adjust, one useful approach to consider may be to gradually phase in "strict" caps over time, by providing for successive reductions in the total number of permits that are available.

As in the case of taxes, the opportunity-cost of using a tradable abstraction/pollution allowance provides both a direct incentive to improve abstraction/avoid pollution and an indirect incentive to innovate for a less-water consumption/pollution intensive future.

The transaction costs associated with some trading systems can be quite high. These costs will affect the net social gains that can be realised from trading. In particular, a requirement for pre-approval of trades stands out as one important barrier; these additional requirements (as well as any other administrative procedures which unnecessarily increase transaction costs) should therefore generally be avoided.

On the other hand, the administrative costs associated with tradable permits systems may be considerably lower than those generated by alternative forms of regulation. A clear distinction can also be made here between cap-and-trade schemes and baseline-and-credit schemes. While the former may have relatively higher start-up costs, they are likely to result in significant savings in terms of running costs over the longer-term.

A key deficiency of baseline-and-credit systems is that the water security cap is not pre-defined. This opens up the possibility that participants in the trading scheme can obtain credit for investments that do little to actually improve water security. As a result, the "Business-as-Usual baseline" – the point beyond which credits begin to be earned – deserves particular attention when designing these programmes. Given that good information about costs and technological opportunities to improve water security is more likely to be available to the regulated sources than it is to the public authorities, there is a danger that the baseline may be defined in such a way that the abstractors/polluters will eventually obtain credits for investments that largely reflect "business-as-usual" developments.

Unless the water security outcome depends heavily on the level of abstraction/pollution over a particular period, the options of banking and borrowing of permits should be actively considered. These approaches can significantly reduce the economic costs of reaching the desired water security target (even while not fundamentally jeopardising progress toward that target). In turn, this can increase the probability that there will be agreement on even more stringent water security targets over time.

As for water taxes, emission trading systems are better suited to addressing the total amount of a given abstraction/pollution within the geographical area it covers than affecting *where*, *when* or *how* a water consuming/polluting product or practice is being used or implemented. Hence, for water security problems where these latter aspects matter, (as in the case of local water shortage/pollution "hot spots"), a trading system might need to be combined with additional instruments, such as local abstraction/pollution standards.

The method used for the initial allocation of permits in a cap-and-trade system is of great importance for both the perceived fairness of the system and its economic efficiency. Broadly, auctioning the permits to the abstractors/polluters covered by the system is the preferable alternative (rather than distributing them for free to existing abstractors/polluters – this is known as "grandfathering"). Auctioning will raise revenues that – depending on national circumstances – can be used to lower distortionary taxes (thereby increasing economic efficiency) or to increase public expenditures. Auctioning will also limit the realisation of windfall profits for abstractors/polluters that receive the initial credits.

It takes time for permit market participants to become accustomed to trading in the market. At the early stages of policy implementation, this can result in "thin" markets, price volatility, and other phenomena which can undermine the development of the market. Efforts should therefore be made to provide long-term stability for the trading scheme, *inter alia* by announcing the caps that will apply over a relatively long time period.

The problem of market power in trading markets with few participants can often be addressed by broadening the sectoral coverage of the trading system – and (possibly) by broadening the geographical coverage. This will reduce the danger of collusion among existing producers in a given sector – collusion that would seek to keep permit prices high, with the aim to keep potential new entrants out of the market. Using broad sectoral coverage limits this problem, because participants from other sectors have no economic incentive to take part in these illegal activities. Even if there are few sources, market power will not be much of a concern if the initial allocation of allowances is close to the expected final distribution of allowances – or if the allowances are auctioned.

The greatest benefits of tradable permits in the early stages of their implementation may arise from the relaxation of regulatory constraints which have previously been inhibiting the application of simple, but more efficient, technologies which are readily available. Over the longer-term, the price signal emerging from permit trades will provide clear incentives for further innovation and technology development.

Most countries use **public financial support** to encourage water friendly practices and to finance water infrastructure investments. While such support can trigger significant water security improvements, it is important to make sure that it is provided only in cases where public goods are expected to be generated, and to consider whether such support really is the most economically efficient way of reaching a given water security target. In particular, taxing or regulating environmental "bads" will reduce the risk of unintended subsidisation of environmentally harmful alternatives, as well as reducing the need for public funding.

Many different types of public financial support measures (e.g. direct budget allocations or grants; low-interest loans; loan guarantees, preferential tax treatment) are used in OECD countries to promote the achievement of water security objectives and/or the development and diffusion of new water technologies. This financial support is given, *inter alia*, to encourage environmentally friendly practices and to finance large water infrastructure investments which would not be implemented in its absence (e.g. investments in water supply and waste-water treatment). Financial support is also sometimes used in combination with regulation or taxation, in order to ease the burden of regulatees and to facilitate implementation of stricter policy instruments.

Water security objectives may be achieved with several different types of instrument. In general, however, policies that require polluters or users of water to pay for the water security problems they generate are preferable to subsidies. Taxing environmental "bads" – or imposing other types of environmental regulation – can often be a better way of proceeding than supporting environmental "goods", especially when the economy-wide economic costs of financing that support are taken into account. Thus, an important first step in making decisions about public support for water security goals is to carefully consider whether that support is really the most economically efficient way of reaching a given water security target.

When providing support for water services, it is also important to define an appropriate reference level – the level beyond which performance will be considered to have improved. Without this baseline, the public water expenditure programme might be credited with water security improvements that would have happened even if the expenditure programme had not existed. Establishing a baseline level could also facilitate decisions about which polluters (or resource users) actually have the related user/pollution rights, and which ones in particular should receive support for providing the particular environmental benefits that are of interest. Apart from identifying eligible beneficiaries and eligible types of projects, the expenditure programme should have clear objectives and a defined timeframe. When the stated objectives have been achieved, the support programme should be wound up, in order to avoid perpetuating the subsidy beyond what is needed.

Public water expenditure programmes can be relatively complex in their appraisal and selection criteria, or in the administrative rules used to implement them. This can lead to high transaction costs and other forms of economic inefficiency. These criteria should therefore be kept as simple, transparent and direct as possible. The institutions administering the expenditure programmes should also have sufficient capacity to manage them – including the capacity to bear the financial risks some forms of support can involve (e.g. loans and loan guarantees). Neither debt nor, contingent and implicit liabilities (e.g. loan guarantees) should be incurred without explicit and prior approval from fiscal authorities.

Public water expenditure programmes should be consistent with the *Polluter Pays Principle*, with sound public finance principles (e.g. regarding transparency, cost-effectiveness and accountability) and with internationally agreed provisions regarding state aid.

Support programmes should also not have the secondary effect of directly or indirectly encouraging additional demand for, or supply of, water consuming/polluting products or activities in the long-term. For example, subsidies to water-efficient irrigation equipment can increase the irrigated area, leading *inter alia* to increased water shortage and pollution problems.

In order to obtain as many water security improvements as possible for a given amount of available support, it is also useful to consider ways of allocating this support in a way that provides the most benefits to those recipients that are willing to commit to achieving the largest water security improvement per unit of support. Cost-effectiveness analysis, or where justified by project size, cost-benefit analysis, should be used in the selection process. One other way of promoting cost-effectiveness is to use a bidding process in the allocation of the subsidy.

Environmental policy instruments usually operate as part of a **"mix" of instruments** (e.g. several instruments are often applied to the same water security problem). It is the net contribution of the instrument "mix" to social welfare that matters most. The environmental effectiveness and economic efficiency of these mixes can be enhanced by adhering to many of the same principles that guide the use of individual instruments, and by explicitly considering the way in which different instruments interact.

Combining two instruments can sometimes enhance the effectiveness and efficiency of both. For example, a water efficiency labelling scheme can reinforce the benefits that emerge from a water tax, and *vice versa*. To exploit these possibilities for mutual reinforcement, instruments that provide as much flexibility as possible to the targeted groups should be used. Market-based instruments will generally provide this flexibility – but *some types* of regulatory instruments (e.g. ambient-based water security standards) can do so as well.

From the perspectives of both environmental effectiveness and economic efficiency, policy instruments should address a given water security problem as broadly as possible (e.g. covering all sources of pollution in all relevant sectors of the economy). They should also provide similar incentives at the margin to all sources that contribute to the water security problem at hand. Market-based instruments (e.g. emission trading systems and taxes) can provide equal marginal abatement incentives, but this is generally much more difficult to achieve with instruments that do not rely on market-based approaches.

For water security problems that have many dimensions (e.g. water pollution from agricultural sources), it can be appropriate to supplement instruments that address the *total amount* of pollution with instruments that address the *way* a certain product is used, *when* it is used, *where* it is used, etc. In many cases, regulatory instruments, information instruments, training, etc., can be better suited to address these latter dimensions than a tax or an emission trading system. Instrument mixes are also often preferable when direct monitoring of pollution is difficult, as in the case of nutrient run-off from diffuse sources in agriculture.

Except for situations where mutual reinforcement between instruments is likely, or when the instruments address different dimensions of a given problem, the introduction of overlapping instruments should be avoided – because this overlap will tend to reduce the flexibility of target groups to respond in the most effective and efficient manner possible. For example, a standard for manure storage facilities that is applied next to a cap-based nutrient trading system that covers pollution generated from the farming sector would not provide any additional incentives to abate water pollution (at least in the short-term, and as long as the cap is kept constant), but could entail increased costs for the livestock holders.

While abstraction/pollution trading (especially cap-and-trade) systems can provide a degree of certainty as to the water security outcome, the compliance costs that will eventually be faced by abstractors/polluters are likely to be quite uncertain under these systems. This uncertainty can sometimes be reduced by introducing a "safety valve" in the permit prices. In effect, this allows abstractors/polluters to abstract/emit whatever amount they like, in return for paying a fixed price (i.e. a "tax") for any abstraction/emissions for which they do not hold an allowance, should the permit price exceed a pre-defined level. This approach needs to be carefully designed, however, in order to preserve the environmental integrity of the overall abstraction/pollution control system. One way of preserving this integrity would be to require abstractors/polluters who use this "safety valve" to make the necessary abstraction/emission reductions in later years.

It is often preferable to primarily address *non-environmental* market-failures (e.g. market power, incomplete information, incomplete user's rights, split incentives between landlords and tenants) with non-environmental instruments, such as competition policy instruments, improvements to patenting systems, deregulation of the housing markets – rather than using environmental policy instruments to address these problems.

When modifications are made to one part of the instrument mix, the environmental and economic impacts associated with other parts of the mix should also be re-evaluated. This reassessment can be very important when "qualitatively new" instruments are added to the existing mix, such as when a quantity based instrument (e.g. a quota-based trading system) is combined with price-based instruments (e.g. taxes and subsidies). It is also important to regularly review the effectiveness and efficiency of the instrument mix that is in place – to ensure that the programme performance anticipated *ex ante* has indeed been realised.

The principle of matching the level of technical assessment to risk and uncertainty on the resource can be applied to risk management functions, such as types of legal rights of water users and use of policy instruments. As the level of resource use increases along with risks to users, risk management should increase to match the risk and ensure that water management increases in effectiveness. This illustrates the concept that **risk management should match the level of risk**. Where the risks are low, such as when the level of use across a basin is a small fraction of the renewable resource, low levels of management are appropriate and as the pressure increases, management should improve to match the risk.

The level of technical assessment and use of policy instruments could change appropriately as the water risk increases. Table 2.1 provides a framework for matching risk management (in terms of sophistication) to the level of water shortage risk.

Table 2.1. **A framework for matching risk management to the level of water shortage risk**

Risk on resource	Resource assessment	Legal rights of water users	Policy instrument
Low	Simple water balance	Riparian access	Basic access fee
Medium	Uncalibrated model	Prior appropriation	Sale of licences
High	Network of measurements, calibrated model	Tradable water rights	Mature trade

Note: Additional management functions could be added to this risk framework, such as regulation of use, metering of use, and audit and compliance functions.
Source: Prosser (2012).

Pressure on the resource might include factors such as the size and variability of the resource, the number and diversity of uses of water, and the significance of water environments that are at risk.

Different types of legal rights of water users are appropriate with different levels of natural reliability of water and pressure on the resource (Andreen, 2011). Riparian rights to water work well in basins where there are plentiful and reliable supplies of water and demands on the resource are low. They provide regulation of access to water by riverside landholders. At higher levels of use, water user's rights might take the form of prior appropriations of use but as levels of use get higher still, water trading can promote more efficient water allocation. The trade may have to be regulated in some ways to overcome market failures but governance measures that promote a mature and open market will be most effective. This illustrates the point that policy instruments and information about the resource all support each other and need to evolve together as the pressure (or risks) increase.

The appropriate policy response needs to consider not only the current level of pressures on resources, but that these pressures are evolving. In the absence of relative decoupling between pressures on water and GDP growth, pressures on water resources grow over time, and risks to users grow, as populations grow and economies develop. Climate change in the drier parts of the world is also reducing the size of the resource and increasing risks such as drought.

The challenge is to define the levels of risk or pressure on the resource where management needs to change in response to the increasing risk. Changing from one type of risk management, such as legal rights of water users, to another may involve large costs to overcome historical lock-in. This would be particularly important for the system of prior appropriation, which is particularly difficult to reform. Moreover, there is a need to carefully evaluate the circumstances under which increased sophistication of user's rights is justified by the risks posed (Box 2.3).

> ### Box 2.3. **Under what circumstances is the increased sophistication of user's rights justified by the risks posed?**
>
> In Australia, land uses such as plantation forestry, and local farm dams can intercept water resources and reduce river flows that have use rights attached to them, but these intercepting land uses are outside of formal water user's rights. Thus there is a lack of integrity or security in the user's rights which can undermine their value. Through the National Water Initiative (COAG, 2004), Australian governments have undertaken to bring intercepting uses of water inside the user's rights framework in basins by requiring "intercepting users" to have a water access licence where water is already fully allocated.
>
> In Australia, other legal rights of water users such as the extraction of water for livestock and domestic use purposes do not require a licence because of the many users and low level of long established use that does not seem to pose a risk to other users. They have a basic right to use water for that purpose. Land uses other than forestry do not require a licence either so the question is at what level of use is it necessary and effective to govern use through a formal access licence and what can be covered under basic unlicensed user's rights? This raises the issue of low but potentially cumulative risks in terms of gradual depletion of water resources.
>
> A third example is the licensing of water for irrigation. Irrigation water use is measured as the water delivered to the farms but some water drains to groundwater or drains back to rivers providing water for others. Only that water which is evaporated on the farm is truly lost and therefore used. This mismatch between gross and net water use does not pose a problem until the allocated water is traded elsewhere or water use efficiency improves and the savings are used to increase crop area. The gross water used remains the same but the difference between gross and net uses are no longer part of the available resource for others. The solution is to account for net water use (evaporation) not just gross water applied, but the question again is under what circumstances is the increased sophistication of user's rights justified by the risks posed?

The costs of management increase with the levels of risk or pressure on the resource but so too do the benefits obtained from managing water well. If the transition to more sophisticated policy instruments and technical assessment is matched by increasing benefits from water then there are overall benefits to society. If policy instruments or technical assessment lag behind the increasing pressure on the resource, then benefits will be reduced as a result of high uncertainty over the resource, concerns of security of access to water, or economic inefficiency.

In cost-effective risk management the benefits of management should (far) outweigh the costs. Thus the challenge is not to implement the most sophisticated and complete set of risk management measures everywhere but to match the level of risk management (in terms of sophistication) to the level of water risks and circumstances of each country.

Going beyond conventional approaches to address water security

One conventional approach to address water security has consisted of estimating the **investment needs** to reduce water risks. As economic growth proceeds and incomes rise, households and governments increasingly have access to the financial resources needed to reduce losses from water risks. As a result, improved water and sanitation services and increased coverage is typically correlated with GDP (income) growth. The same applies to development of large infrastructure to store water for use during droughts, to supply irrigation schemes, to control floods, and to generate hydroelectric power.

In that context, a minimum platform of traditional physical infrastructure – dams, canals, hydropower plants, irrigation – is seen as a precondition for reaping the economic benefits of water (Briscoe, 2009). As a result, the level of economic development should determine risk management policies. The argument follows that in non-OECD countries, the priority should be to mobilise financing to build the infrastructure platform necessary to achieve water security objectives. In contrast, the priority in OECD countries should be to create incentives for rational allocation and efficient use of water.

Another approach to address water security has been to assess the costs and benefits of **"closing the water supply/demand gap"** over a given time horizon (e.g. by 2030), typically based on marginal cost curves. Marginal cost curves aim at identifying the most cost-effective (including innovative) technologies to close the gap (e.g. reuse of treated wastewater, drip irrigation). The range of (market and non-market) benefits associated with closing the water gap is also assessed (2030 WRG, 2009).

These approaches have significant shortcomings. For instance, the approach that aims at sequencing actions according to the level of country development compares the costs likely to be incurred by a hazard event (water shortage, excess or pollution) with the costs of changing the probability of that event through structural interventions (larger reservoir and bulk transport systems, higher flood defences, more advanced water and waste water treatment plants). But it does not pay attention to reducing the consequences arising from the hazard event by altering the vulnerability of the potentially affected populations, ecosystems or physical assets, as would be the case of a risk-based approach. It also overlooks the question of what is an acceptable or tolerable level of water risks for various water users in a given society. This needs to be considered in light of the costs of managing water risks, as well as in light of competing priorities. Moreover, a risk-based approach would allow the identification of areas of high risk where policy action should be given priority, and this applies for both OECD and non-OECD countries.

The marginal cost curve methodology seeks to manage water for certainty. It provides a static "snapshot" of the cost of various interventions at a given point in time and fails to capture the complex interconnections of water resources supply and demand. In contrast, a risk-based approach would seek to set acceptable levels of water risks. Risk management would go much beyond a simple (generic) assessment of technical measures to manage water risks. It would look at a range of policy instruments (including technology) and assess how much each instrument or instrument mix is likely to best contribute to the water security objectives from an economic efficiency and equity perspective.

Country cases

Case studies presented at an OECD expert workshop convened in 2012[1] show that there are already some elements of a risk-based approach in the way OECD countries currently address water security. This section provides real-world examples of setting targets for water risks and implementing policy instruments to achieve the targets. The country cases include the Murray-Darling Basin (Australia),[2] Alberta's South Saskatchewan River Basin (Canada),[3] France,[4] England and Wales[5] and the Central and West Coast Basins of Los Angeles (United States).[6]

Managing risks of water shortage

Two risk management strategies have been the central focus in addressing the risks of water shortage in **Australia's Murray-Darling Basin** (MDB), which is characterised by an extreme variability of river inflows. First, the introduction of water markets created incentives for more efficient use of water in agriculture. Second, attempts are underway to reduce consumptive water use to levels considered to be environmentally sustainable.

The *risks of irrigation water shortage* resulting from over-allocation of irrigation water led authorities to close the MDB to the allocation of new water user's rights. In addition, a cap on water use was introduced in 1995 to limit further extraction of water from the MDB at 1993-94 levels. The cap did not limit development; it prevented increases in water use, requiring water efficiency measures to be the main driver of productivity gains. This was the first essential step in introducing water markets into the MDB by imposing a level of scarcity on the water user's right.

The development and expansion of a functioning water market in the MDB took place over more than a decade, between 1994 and 2006. The Council of Australian Governments (COAG) Water Reform Framework (1994) was instrumental in allowing the water user's right to be unbundled from the right to use and own land,[7] and implement a water market with tradeable water user's rights. By 1996, trade within a State was allowed, and two years later, a trial of interstate trade had begun along the borders of South Australia, New South Wales and Victoria. By 2004, with the introduction of the National Water Initiative (NWI), the barriers to entry were reduced to further encourage trade. Full interstate water trade in the southern connected MDB began in 2006. Another important action undertaken by states and territories in response to the NWI was the unbundling of water from land. This action permitted water to be transferred independently of land between users without the need for consideration being given to various land use requirements.

While there are trade-offs between water demand from cities and for energy production, the volume of water involved is minor compared with the requirements for environmental flows and irrigation in the Basin. Yet, the Basin is home to over two million

Australians, with another 1.3 million partially dependent on the Basin for their water supply. The Australian capital city is within the Basin and another state capital, Adelaide, currently draws up to 150 GL of water from the River Murray. Since 2008, Ballarat has used 33 GL of water from the MDB, and the Sugarloaf pipeline can deliver 75 GL of MDB water to metropolitan Melbourne in case of critical human needs. However, it is unlikely the Sugarloaf pipeline will be used in the medium term, as Melbourne's long term water security is assured as a result of its newly built desalination plant. Similarly, Adelaide's demand has diminished following the construction of a desalination plant with the capacity to supply 100 GL per annum. In total about 2% of consumptive water in the MDB would meet current urban demand. The volume of water required to meet domestic demand is very small and has not resulted in major trade-offs between cities and agriculture.

The 2007-08 drought emphasised the success of water trading in maximising the productive output of irrigated agriculture. Through the realisation of the opportunity cost of water as a scarce and valuable resource, water markets allowed flexibility in water use. Water trading allowed marginal irrigators to exit the industry or to restructure their business model more profitably, by realising the value of their water rights. Markets also increased the movement of water into South Australia, where horticulture dominates irrigated water use, reducing the salinity levels from what they would have been without water trading. The experience of severe drought in the MDB showed that water markets support irrigation productivity, help farmers manage risk and adjust to seasonal variation, provide a secondary source of income in the event of crop failure, increase the uptake of sophisticated farm management practices, assist the process of financial restructuring, encourage structural change in the irrigation sector, maintain high-value permanent plantations through periods of water shortage, and increase conveyance flow to downstream users with some environmental benefits (Fargher and Olszak, 2011).

Water availability to "general security" license holders is announced as a proportion of entitlement, commonly referred to as an "allocation". The announced allocation depends upon the resources currently available in storage and those resources expected to be available during the season. "High-security" water entitlements generally have all of their allocated water delivered each year even though they are also subject to announced allocations (Rowan et al., 2006). Allowing for several types of water entitlements with varying levels of security allow users to express different risk preferences reflected in the price of entitlements.

The establishment of water markets also provided a mechanism for governments to address the *risks of environmental water shortage* through reallocating water between consumptive and non-consumptive use. In 2002, COAG agreed to take a first step in reallocating water to the environment, through the Living Murray Initiative. This decision was based on an increasing realisation, since the cap on development in 1995, that a balance had to be struck between the economic and social benefits derived from developing water resources in the MDB, and the environmental uses of water in rivers. The *Water Act (2007)* went further than the Living Murray Initiative in trying to rebalance water use between irrigators and the environment. It required the establishment of two new institutions: the Murray-Darling Basin Authority (MDBA) and the Commonwealth Environmental Water Holder (CEWH). The MDBA was tasked with developing a Basin Plan that defines the sustainable diversion limit for each river and groundwater management area in the MDB for implementation in 2019. The Basin Plan was approved by Parliament in November 2012.

In addition to measures to meet basic environmental flow requirements, the government is acquiring water user's rights with the objective of returning more water to the environment. This water is held by states and the CEWH, a central agency that is given independent control over how to deliver water to maximise the benefits for basin-level environmental objectives. The CEWH does not have special privileges within the water market. It abides by the same legislative rules and regulations as other water user's right holders, and the water entitlements it holds have the same characteristics as it would if otherwise held by a private water user's right holder. Environmental managers have the same user's rights regarding their water entitlements and allocations as other water user's right holders (i.e. they can transfer, buy, sell, store and carry-over water).[8]

In this way, management of environmental flows in the MDB has changed from water delivery based on regulatory measures, to one based on both regulatory measures and equal entitlements. While this is a significant change in focus, the vast majority of water provided to the environment is still provided through regulatory measures in water sharing plans. Such management of environmental flows provides a pool of water that can be used to deliver targeted environmental flows in a more dynamic and flexible manner than before. Flows can be used to reintroduce, for example, the flood pulses that have been eliminated from the system as a result of extensive regulation, thereby reducing the accumulation of salinity, nutrients and pollutants in wetlands and river channels. The flood pulses should also trigger breeding of fish and provide other benefits to aquatic and riparian ecosystems.

In essence, water held by the CEWH can be delivered for environmental benefits according to basin-wide objectives. The recently adopted Basin Plan sets the amount of environmental water that needs to be recovered in order to meet sustainable diversion limits. In 2008 nearly AUD 9 (USD 7.4) billion had been provided to recover water for the environment through increased irrigation efficiency (e.g. on-farm irrigation upgrades, upgraded water delivery infrastructure and metering) (65%) and buybacks of water user's rights (35%). In October 2012, the Prime Minister of Australia committed a further AUD 1.775 (USD 1.837) billion for a new programme, to begin in 2019, that will also address key system operational constraints currently restricting the regulated delivery of environmental water throughout the MDB (DSEWPaC, 2012a,b). Around 4 000 GL/yr should be recovered by 2025 (Figure 2.2), compared to the long-term average water use in irrigation of 11 000 GL/yr.[9]

The Government of Alberta has also taken steps to manage risks of water shortage in **Alberta's South Saskatchewan River Basin** (SSRB). A dense (and growing) population, the effects of climate change, an existing arid climate in much of the basin, and growing demand among competing users – coupled with the need to ensure sufficient water to protect aquatic and ecosystem health – all place significant pressure on the basin. Further compounding the challenge is the requirement that 50% of the annual natural flow must pass to Saskatchewan, a neighbouring province.

To address the risk of water shortage, the South Saskatchewan River Basin has been closed to new water allocations. In addition, the province evolved its water allocation policies and the associated legislation, culminating in the proclamation of a new *Water Act* in 1999. Pursuant to the Water Act, the right to divert and use water is granted by a licence or registration under the *Water Act*. The *Act* requires that a licence be obtained before diverting and using surface water or groundwater, except for household or domestic use, traditional (non-irrigated) agriculture, fire suppression, and other small-quantity uses by riparian landholders. Licences identify water sources, points of diversion, maximum

Figure 2.2. **Environmental water to be recovered in the Murray-Darling Basin, 2006-24**

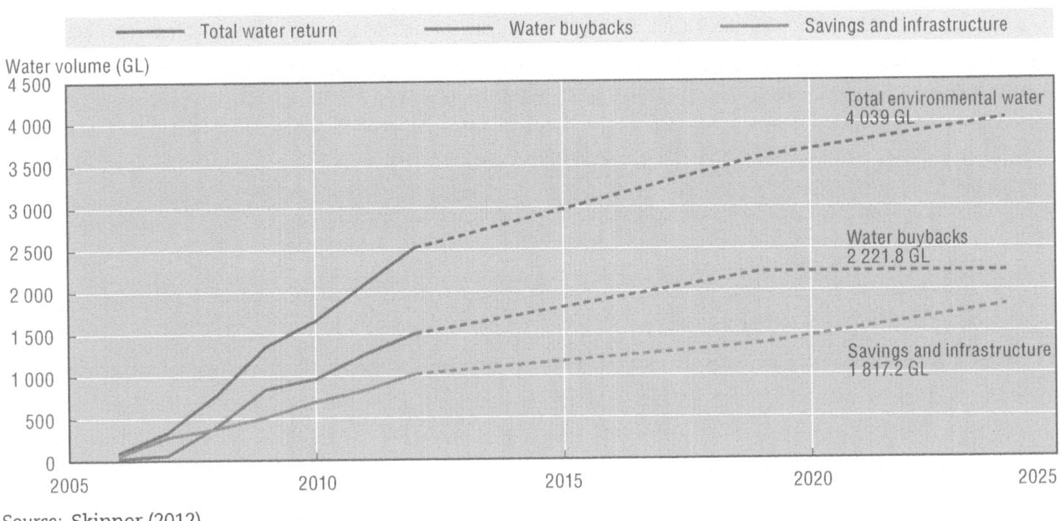

Source: Skinner (2012).

allocations, the purpose of the projects, rates and conditions of diversion or withdrawal, and the priorities of the water user's right.

The licencee is given an annual allocation, which is the maximum amount of water the user is allowed to divert each year. The licence also provides an estimate of consumptive use, losses and return flow. Though these estimates are not enforceable, they provide sufficient information on which to base the annual licenced use (the annual allocation less the return flow) for the project at the time the application was made. Many licencees don't use all the entitled water in their licence. In particular, actual use for irrigation licences varies from year to year, depending on such factors as weather, economic conditions and crop rotations.

Licence applications are reviewed for impacts on the source water, the aquatic environment, public safety, and on other users. Under the Environment Protection and Enhancement Act, approvals are required for activities with a high potential to impair or damage the environment, personal property, human health, and safety. Environmental Impact Assessments are mandatory for dams more than 15.0 m high, for diversion structures and canals with flow capacities greater than 15.0 m^3 per second and for reservoirs with a storage capacity greater than 30 000 dam^3 (30 million m^3).

All licences are given a priority number, based on the date a completed application is received by the Government of Alberta. Higher priority (earlier) projects are entitled to their full water requirements before projects with lower priorities can divert water. When streamflow and demand data indicate a trend toward deficits, the operations and management of provincially owned storage projects are reviewed as alternatives to imposing diversion restrictions. Licencees, in accordance with their priority dates, may be directed to stop diverting water until the minimum stream flow has been restored and the needs of all higher priority licences can be met.

Pursuant to the Water Act, the province created new rules to allow transfers of water user's licenses within the basin – with each voluntary transfer returning 10% of the volume to the river, which serves to enhance and support protection of the aquatic ecosystem (Box 2.4).

> **Box 2.4. Water allocation transfers in Alberta's South Saskatchewan River Basin**
>
> In a basin closed to new licences, Alberta's *Water Act* allows water allocation transfers to move water between users if the transfer is consistent with an approved water management plan. Enabling transfers in a basin allows water users the flexibility to address their water needs. Transfers can also benefit the aquatic ecosystem if some portion of the transfer is retained for environmental purposes. The transfer is limited to the *user's right* to divert a volume of water from a source of water supply, under a certain priority. There is no physical transfer of water from the land. This type of transfer is voluntary, with a willing seller and willing buyer. Transfers can be permanent or temporary. With a temporary transfer, the transferred allocation reverts to the original licencee after a specified time period.
>
> The Alberta Government monitors this system through a number of control mechanisms. These require that a transfer must first be authorised in a water management plan or through an order of Cabinet. When a transfer is reviewed by the Alberta government, it can choose to withhold a percentage (up to 10) of the transferred water to be held in a Water Conservation Objective licence that is not available for reallocation for other uses.

In **England and Wales**, the "Water for Life" programme (Water White Paper 2011) sets out an ambitious agenda for reforms, including considering options to increase flexibility through water trading. Indeed there have been droughts in one or more regions of England and Wales in 12 years out of the past 22 – droughts are an important driver of policy reform and raise the issue of how to address the risks of water shortage. The potential need for more water for irrigation is included into policy scenario analyses. Understanding/ quantifying the water needed for the environment is a focus and the idea of ecosystem services is gaining ground.

The Environment Agency for England and Wales has published for every catchment, reports setting out the total resource, the environmental allocation, the volumes currently allocated, and the volumes remaining (if any) at different flow states. These Catchment Abstraction Management Strategies (CAMS) provide information to support water trading, although little has so far taken place.

One of the biggest pressures on water resources is the projected population growth. By the 2030s, the total population of England and Wales is expected to grow by an extra 9.6 million people. The Agency has looked at the potential effect growth, societal change and climate will have on future demand. By 2050, according to different scenarios, total water demand is expected to vary from 15% less than today to 35% more.

Two conclusions are evident from the Agency's assessment of current water resources availability, and the assessment of future availability. The first is that the legacy of over-abstraction, inherited from predecessor bodies, needs to be tackled in order to ensure a sustainable baseline for the future. The second conclusion is that the existing system of resource allocation, which the Agency oversees, needs reform to make it more flexible and dynamic in the face of future uncertainty.

The system of water allocation, through abstraction licences, has been in operation since 1965. Pressure on water resources is now at the point where there is very little water availability for new licences during the summer months, and even winter licences may be constrained by flow conditions to protect the environment, and the user's rights of existing abstractors (a fundamental tenet of the legislation is to prevent derogation). All new licences are time limited and subject to conditions to protect the environment.

Water has been allocated historically, and continues to be so, on the basis of "first come, first served". There is no legislative or policy hierarchy of use. However, legislation does allow irrigation to be curtailed to protect the environment during a drought, and to recognise the primacy of public water supply and the need to protect human health during periods of extreme water scarcity. At such times, special powers allow the environment and other water users to be impacted in order to minimise the risk of supply failure. However, environmental mitigation, and financial compensation, are necessary compromises. The Agency oversees drought planning and can grant drought permits to water companies to allow them to operate outside normal limits during a drought.

Some trading in abstraction licences already takes place, almost entirely between farmers for irrigation. However, it is recognised that the current resource allocation system does not reflect the value of water or its scarcity, nor allocate it to the highest value use. Current allocations are largely fixed, with few real signals to abstractors, or indication of the value of ecosystems.

The government is committed to reform, to develop a more market-based approach to water allocation, and a system which is more flexible and adaptable to future demands and uncertainty. And which provides greater protection to the environment. But before this can be implemented, there needs to be a sustainable baseline for abstraction, to avoid the problems of markets developing in unsustainable abstraction rights seen elsewhere in the world.

Tackling the legacy of over-abstraction is not straightforward. At the moment, the damage it causes to the environment is not really reflected in the price paid for water. The government supports the use of charges levied on abstractors to address unsustainable abstraction,[10] and the Agency is piloting reverse auctions as a mechanism to claw back abstraction rights. It is also removing real and perceived barriers to trading, and will identify catchments with potential for increased trading to test a reformed abstraction regime.

The Agency's analysis of future water availability demonstrates the importance of increasing interconnection in water supply systems. Tackling supply deficits within individual catchments will be a high cost approach, increasing the requirement for new infrastructure, and requiring more constraint on water use. However, water is heavy and pumping – and carbon – costs are high, so large scale, long distance transfers are expensive, relative to the water's value. But there is scope for greater interconnection within and between water companies. The English government is looking for Ofwat (the Water Services Regulation Authority) to use incentives to support bulk transfers and interconnections.

In **France**, "water apportionment zones" were introduced in 1994 to address the *risks of irrigation water shortage*. The aim is to delineate areas of chronic surface water or groundwater deficit (i.e. water supply insufficient to meet demand) (Figure 2.3). These areas are subject to more stringent abstraction licensing and higher abstraction charge. An estimate of abstractable amounts (supply) and uses (demand) must be made to forestall the risk of shortage (and define priorities for use). These estimates, now being carried out in river basins, will help improve the system of water apportionment zones and abstraction licensing.

For each water apportionment zone, the aim is to set new abstraction limits to reduce the risk of water shortage to two years in ten, while allowing for flexibility to reduce the limits if weather conditions so require. Allocation priority will be given to drinking water and ecosystems. In areas where the deficit is particularly liable to affect agriculture, the

Figure 2.3. **Water apportionment zones for surface water and groundwater, France**

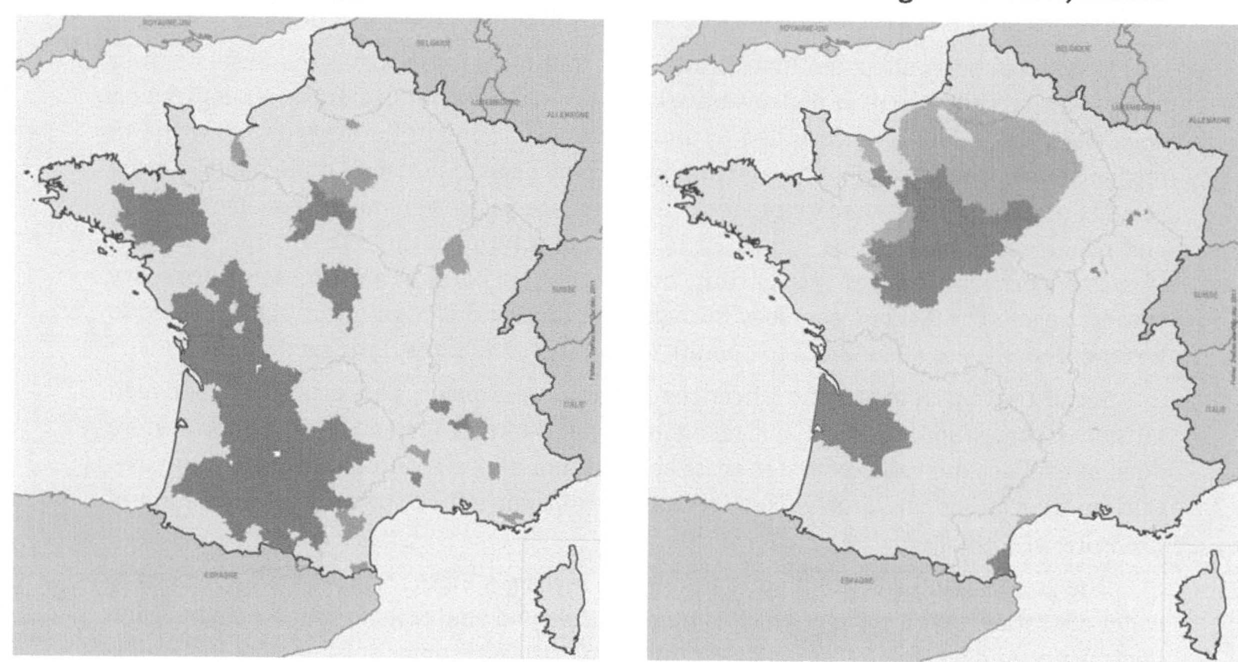

Notes: Dark: zones classified in 2009, Light: zones classified in 2010. Since 2010, some zones have been declassified and new zones classified as water apportionment zones. Left: Surface water; Right: Groundwater.
Source: Ministry of Ecology, *Sustainable Development and Energy* (Water and Biodiversity Directorate), in Ben Maïd (2012).

new system entails the setting of an abstraction limit for the whole farming sector (via issuance of abstraction licenses to irrigator associations and chambers of agriculture) and allocation of individual (non tradable) water quotas to farmers within the limit. Irrigator associations and chambers of agriculture will manage the trade-offs between farmers within the quota allocated to agriculture,[11] report on actual withdrawals and manage quota reductions in case of drought. There is no plan to introduce tradable irrigation water quotas in the short term. In a major wheat producing region (Champigny aquifer), an experimental system allocates irrigation water quotas based on crop rotation but without yet a cap on total irrigation water use.

When addressing the *risks of environmental water shortage*, water needs for ecosystem use are complex to define. The aim is to ensure the survival of fauna and flora in the environment (and hence appropriate physical, chemical, volume and temperature conditions), the long-term future of habitats (and hence the hydromorphology, which depends on flow and not just volume) and the ecosystems' resilience to meteorological or climatic disturbances and to abstractions for anthropic use.

Faced with these difficulties of defining a minimum volume or flow specific to each ecosystem, France has decided that a flat-rate proportion (between 10 and 20%) of the flow – defined as a standard (*"débit quinquennal sec"* – QMNA5)[12] – should be set aside as the minimum biological flow. This flow is a minimum. New studies may lead to set the minimum flow at a higher level provided acceptance by users in basins.

A major concern in the **Central and West Coast Basins of Los Angeles (United States)** is preventing water shortage risks in case of earthquake. Water delivery to the coastal Los Angeles area could be disrupted by a range of events. A recent study estimated the likely impacts of a large scale, but plausible, earthquake on the San Andreas Fault in southern

California (Jones et al., 2008). Disruptions of imported water conveyance and basin-wide distribution were determined to be major impacts. A follow-up analysis of water system impacts (Davis and O'Rourke, 2011) estimated that it would take at least 12-15 months for water service to be completely restored for much of the affected population, and that the costs of business interruption losses due to water supply reduction would likely be much greater than USD 50 billion. The study concluded that there is a need to increase local water storage. Other events that could disrupt delivery of imported surface water include a breach in the levees in the Sacramento-San Joaquin Delta in northern California and extreme droughts. Extreme droughts would also limit the capture and reuse of local runoff and could impair the quality of imported water.

The US Geological Survey (USGS) conducted a study to evaluate the role that groundwater can play in water-supply emergency planning in the coastal Los Angeles region (Reichard, Li and Hermans, 2010). The USGS used a groundwater simulation model to generate response functions representing the basin-wide hydraulic impacts of different scenarios of disruptions and utilisation of groundwater during emergencies. These response functions were coupled with cost coefficients, a discount rate, and a probabilistic representation of the likely additional groundwater use to estimate the emergency benefits of groundwater management strategies. The analysis focused on artificial recharge, but could also be applied to strategies for increased conservation and expanded use of recycled water.

While there are many simplifications and assumptions in the approach (including a very simple representation of subsidence and the assumption that there is available well capacity and an intact local distribution system), our analysis confirms that groundwater can, in fact, provide an important component of increased local storage during emergencies. It is possible to quantify the benefits of artificial recharge for increasing groundwater available for emergency use. These benefits correspond to reduced negative impacts (e.g. greater pumping lifts, land subsidence, and seawater intrusion) when additional groundwater is utilised during an emergency. For the application to the Central and West Coast Basins, expected emergency benefits of artificial recharge are dominated by reduction of potential subsidence costs (Figure 2.4).

The analysis could be expanded to explicitly incorporate well-capacity constraints and consider potential earthquake damage to the local water distribution system. The extent of damage to the distribution system, and the time and cost required to make necessary repairs could be represented probabilistically. These extensions of the analysis would allow assessment of the benefits of investing in additional well capacity for emergency supply purposes and improvements in the resiliency of the water distribution system.

Innovative strategies for conjunctive management of groundwater and surface water are required to ensure sustainable, resilient water supplies throughout the United States. The approach demonstrated in coastal Los Angeles could be applied to assess the emergency benefits of these conjunctive use strategies.

To **sum up**, in both the Murray-Darling Basin (Australia) and South Saskatchewan River Basin (Canada) over-allocation of irrigation water led authorities to close the basin to the allocation of new water user's rights. In addition, in the Murray-Darling Basin, limits have been set on diversions from rivers and abstractions from groundwater. The limits can change from year to year, providing scope for greater water use in certain years and lower water use in other years.

Figure 2.4. **Benefits of artificial groundwater recharge for emergency use, Central and West Coasts Basins of Los Angeles**

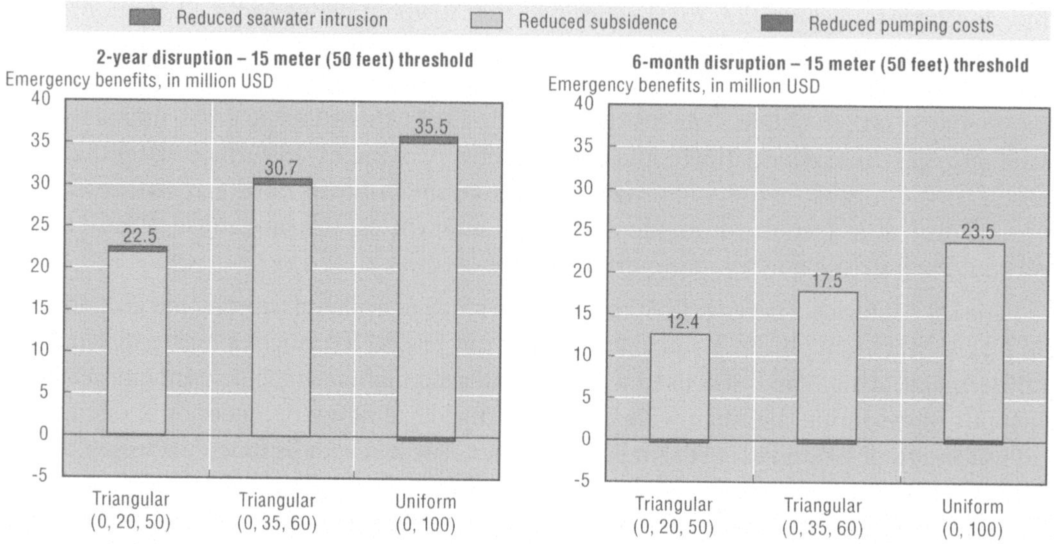

Note: "triangular" and "uniform" refer to the % of imported water replaced by groundwater. Triangular (0, 20, 50) implies a minimum % of 0, a maximum % of 50, and a most likely % of 20. Uniform (0, 100) implies that % from 0 to 100 are equally likely. These probability distributions assume progressively greater likelihood of imported-water replaced by groundwater.

Source: Reichard, Li and Hermans (2010).

In these two basins the level of risk management matches the level of water risk. Because of the high risk of water shortage in the Murray-Darling Basin, Australia undertook reforms to implement tradable water user's rights. Instead, because the risk of shortage is not as acute in the South Saskatchewan River Basin, Alberta only implements licencing of annual allocations.

As is now the case in Alberta, water in England and Wales has traditionally been allocated through abstraction licencing. The recent drought events, though, have prompted the UK government to consider introducing water trading in England and Wales. In France, an experiment in the Champigny aquifer (a major wheat producing region) introduced (non tradable) irrigation water quotas, based on crop rotation. The next step will be to set a cap on total irrigation water use, with irrigator associations allocating individual (non tradable) quotas within the cap.

To better target the risks of water shortage, France is improving the delineation of chronic water deficit areas (these areas are subject to more stringent abstraction licensing and higher abstraction tax). Compared with Australia's Murray-Darling Basin, where the target (diversion limits) aims to prevent increases in diversions above 1993-94 levels, in France the target is set to meet water demand in eight years in ten.

In Alberta, licencees may be directed to stop diverting water until the minimum stream flow has been restored. In Australia, in addition to measures to meet basic environmental flow requirements, the government is acquiring water user's rights to ensure that more water is returned to the environment. In France, a new water allocation policy gives priority to drinking water and ecosystems; sectoral needs come next.

In the United States, coastal Los Angeles is known to be at risk for earthquakes. The region relies on imported surface water, which delivery could be disrupted in case of earthquake. Managing such water shortage risk requires additional reliance on the region's groundwater while preventing land subsidence associated with groundwater depletion, which entails implementing groundwater recharge. The target for the risk of water shortage (i.e. the share of imported water replaced by groundwater in case of earthquake) can be determined by the reduction of expected land subsidence costs.

Dealing with inadequate water quality

In **European Union (EU) countries**, targeting the risk of inadequate water quality translates into looking at general and more specific statutory water quality requirements. EU water quality objectives require ecological and chemical protection everywhere as a minimum. These two general requirements are referred to as "good ecological status" – defined in terms of the quality of the biological community, the hydrological characteristics and the chemical characteristics – and "good chemical status" – defined in terms of compliance with all the quality standards established for chemical substances at EU level. More stringent requirements are needed for particular uses, such as specific protection of unique and valuable wetland habitats, protection of drinking water resources, and protection of bathing water, for which specific protection zones must be designated.

The EU addresses trade-offs between the risk of inadequate quality and other water risks (e.g. flood risk, risk of shortage) through providing derogations from the requirement to achieve good status. This is the case of uses which adversely affect the status of water but which are considered essential on their own terms – they are overriding policy objectives. The key examples are flood protection and essential drinking water supply. Derogations are provided so long as all appropriate mitigation measures are taken.

EU risk management policy for groundwater is different. General protection calls upon the precautionary approach. It comprises a prohibition on direct discharges to groundwater, and (to cover indirect discharges) a requirement to monitor groundwater bodies so as to detect changes in chemical composition, and to reverse any antropogenically induced upward pollution trend. The presumption in relation to groundwater is that it should not be polluted at all. For this reason, setting chemical quality standards are not seen as the best approach, as it would give the impression of an allowed level of pollution to which EU member states can fill up. A very few such standards have been established at EU level for particular issues (nitrates, pesticides and biocides).

In **France**, increasing priority is given to preventive actions over curative actions in addressing the risk of inadequate water quality. This is the case of drinking water, where the protection of water sources is seen as more cost-effective than end-of-pipe water treatment. For example, starting in 1996, the municipality of Lons-le-Saunier (a town of 20 000 inhabitants) gradually introduced financial aid packages for farmers within a perimeter of 270 hectares (667 acres) of drinking water abstraction points. The aim is to encourage them to stop growing maize, make less use of plant protection products, stop using certain products, leave grassed strips, and cover the soil. The cost of such support is only EUR 0.01 per m^3 of water distributed (Agence de l'eau Seine-Normandie, 2009).

Beyond such local contractual arrangements to encourage them reduce the use of farm inputs,[13] farmers are subject to a pesticide tax. The tax, though, does not apply to fertilisers.

Pollution charges apply to direct discharges to water from large industrial plants (including large-scale livestock units) based on actual amount discharged. When connected to public sewerage, industry must comply with quality standards for wastewater discharges into sewerage networks.

For households, as it would be very costly to monitor (meter) individuals' emissions, pollution charges are based on estimated amount discharged.

In **Australia's Murray-Darling Basin** (MDB), to address the risks of in-stream salinisation, salinity targets were set and the salinity impacts of agricultural development managed through a credit and debit scheme and joint government investment in salt interception schemes. If investment in salt interception schemes was entered into, and salinity credits could be created, then irrigation use could continue to expand. Some states encouraged irrigation in areas of minimal impact to river salinity through the zoning of salinity impacts.

More precisely, the Salinity and Drainage Strategy (1988) entails: 1) establishing end-of-valley salinity targets for each tributary catchment and a Basin target – less than 800 Electrical Conductivity (EC) units for 95% of the time – at Morgan in South Australia; 2) developing a system of salinity credits and debits with states which link these (non tradeable) credits and debits to the level of development of agricultural land, providing additional "salt disposal entitlements" for states to offset new irrigation development; 3) developing salt interception schemes – large scale groundwater pumping and drainage projects that intercept saline water flows and dispose of them into drainage lagoons where evaporation takes place and salt crystals can be harvested as a product for sale; 4) some MDB states introduced salinity impact zones which facilitated state accountability of salinity impacts of agricultural development, including offsetting mechanisms. The salinity zoning system provides a clear set of planning rules when establishing new irrigation development or relocating water from one part of the catchment to another.

The Salinity and Drainage Strategy has been successful in moving water use from highly saline areas where old irrigation technology was being used (e.g. the Kerang area) to areas where new irrigation standards had to be met to ensure minimal salinity impact (e.g. the Mallee area). The water market (see above) was extremely helpful in assisting the rebalancing of water use, and was the driver in establishing the whole salinity zoning system in Victoria.

Managing flood risks

In **France**, introduced in 2002, programmes of action to prevent flood risks seek to promote integrated management of flood risks at basin level through contracts between central and local government. More locally, natural risks prevention plans regulate land use according to the natural risks to which the land is exposed. The plans identify high-risk zones on the basis of historical and scientific analysis of local phenomena in consultation with local stakeholders. They define measures to reduce the exposure to risk, such as the prohibition or restriction of construction, which may apply to existing property. The natural risks prevention plans are attached to territorial planning documents and are linked in regulatory terms to the CatNat system described below (Letrémy and Peinturier, 2010).

In addition, under EU Directive 2007/60/CE on the assessment and management of flood risks, France must make a preliminary assessment of flood risks. The assessment may include an evaluation of the economic impact of floods, which may contain a cost-benefit analysis and a multi-criteria analysis of the effectiveness of flood prevention measures.

The Barnier fund was created in 1995 to finance the expropriation of properties exposed to natural risks that presented a grave threat to human life (compensation, measures to secure sites). It is funded by insurance companies through a levy on the product of premiums and additional contributions relating to the natural disaster cover (CatNat system) included in insurance policies. The system is unusual in that all insurance policy holders (home insurance, vehicle insurance) help to fund it.[14] The levy was gradually increased, from 4% to 12% for home insurance, to match the rise in spending on natural risk prevention, especially against flooding. It now raises some EUR 150 million a year.

Securing resilience of groundwater systems

The historic development of the Los Angeles area has been very closely tied to water. It led to the extensive depletion of groundwater in the **Central and West Coast Basins of Los Angeles** during the first half of the 20th century.

Multiple sources of water provide the water supply to the four million people in the Central and West Coast Basins. About 308 million m^3/yr (250 000 acre-ft/yr [af/yr]) are pumped from groundwater, and about 382 million m^3/yr (310 000 af/yr) are provided by imported surface water. In addition to the direct delivery of imported surface water, about 148 million m^3/yr (120 000 af/yr) of additional surface water (imported, recycled, and local runoff) are applied in spreading ponds, and about 37 million m^3/yr (30 000 af/yr) of imported and recycled water are injected for control of seawater intrusion. Imported water comes from three sources: the California State Water Project, Owens Valley, and the Colorado River. Most of the water purveyors in the Central and West Coast Basins have the ability to use both imported water and groundwater.

Groundwater depletion and the increasing risks of undermining resilience of a groundwater system led to the formation of the Water Replenishment District of Southern California (WRD) in 1959. WRD has authority to collect an assessment from all pumpers in order to purchase and recharge water for artificial replenishment of the groundwater system and makeup the annual overdraft.

Groundwater depletion also led to the legal adjudication of the basins in 1961 and 1965, respectively. The adjudications included the establishment of water user's rights that set limits on the amount of water that can be withdrawn by each groundwater pumper.[15] The total amount of pumping rights is fixed, but user's rights can be bought, sold, or leased to any party on the open market (cap-and-trade system).

In **EU countries**, targeting the trade-offs between risk of water shortage for consumptive uses and risk of undermining the resilience of groundwater systems translates into looking at groundwater abstraction limits. Groundwater abstraction is limited by law to the portion of the recharge that is not needed by the ecology (so-called sustainable resource). Indeed some of the recharge is needed to support connected ecosystems (whether they be surface water bodies, or terrestrial systems such as wetlands).

Applying a water risk strategy with a basin approach

Water risk management in the **South Saskatchewan River Basin** is being addressed through a number of comprehensive, adaptive, and innovative ways, including a water strategy and a land-use framework. The government stresses that all in the basin are stewards of Alberta's water resources. There are shared responsibilities among the government, citizens, industry, municipalities, and stakeholders – and these groups together must discuss, debate, and consider what the most appropriate social, environmental and economic outcomes for the basin should be. Implementing market-based instruments and the valuation of ecosystem services are further challenges ahead.

A new overarching strategy for water risk management – *Water for Life: Alberta's Strategy for Sustainability* – was released in 2003. *Water for Life* is the overarching government-wide strategy for water in Alberta. It represents a shift from a government-centred, regulatory approach that focuses on water allocation, to one that incorporates principles of place-based management, watershed management, and a shared responsibility for the stewardship of resources.

The strategy guides policy measures across Government of Alberta ministries. It was based on three major goals: safe, secure drinking water supplies; healthy aquatic ecosystems; and reliable, quality water supplies for a sustainable economy. *Water for Life* identifies the need to engage all Albertans in managing Alberta's watersheds. This translates into the Government of Alberta working with other governments, stakeholders and the public to collaborate in three types of partnerships to share responsibility for identifying solutions to watershed issues in Alberta. The Alberta Water Council, a multi-stakeholder advisory body, provides overall guidance on the implementation of the strategy. Regional organizations, the Watershed Planning and Advisory Councils, raise awareness of the state of Alberta's watersheds, and act as catalysts in creating, implementing, and assessing long-term watershed management plans. Similarly, Watershed Stewardship Groups undertake "on-the-ground" activities to protect the health of the basin.

Alberta's prosperity has created opportunities for the economy and people, but it also has created challenges for Alberta's landscapes. There are more people doing more activities on the same piece of land. To address these concerns and manage the impact of land use on water risks, in 2008, the Government of Alberta introduced the *Land-use Framework*. The framework sets out an approach to manage public and private lands and natural resources to achieve Alberta's long-term economic, environmental and social goals. It provides a blueprint for land-use management and decision-making that addresses Alberta's growth pressures.

The framework complements the province's *Water for Life* strategy. Indeed, what uses are permitted on land – or more precisely, how they are done – clearly impact adjacent watersheds. Implementation of the *Land-use Framework* is a key vehicle for implementing Alberta's "cumulative effects" management system at the regional scale.

Alberta's shift to a cumulative effects management system is more effective and efficient, and seeks to consider the cumulative effects of all activities on a watershed. The current system is evolving and adapting to "place-based challenges", which allows decision makers to see the big picture and help those on the landscape to be more strategic and responsible in their development activities. Within this system, various tools, resources and relationships work together to comprehensively manage activities that affect the environment, economy and society in a particular place. It is an adaptive management system that follows the approach of setting, meeting and evaluating place-based outcomes (Table 2.2).

France has a long tradition of integrated water management at basin level, incorporating all the risks exerted on the resource.[16] One of the features of the management system is the systematic use of participatory democracy as a form of governance, notably through river basin committees. This system for permanent consultation of and negotiation with stakeholders is certainly one of the conditions for ensuring that water management measures are socially acceptable. The time taken to reach decisions inherent in negotiation processes may be open to criticism, but in an emergency, central government and local

Table 2.2. **Key developments in managing water risks in Alberta's South Saskatchewan River Basin**

From (pre-2003)	To (water for life, 2003)	To (land-use framework, today)
Paradigm of abundance of natural resources.	Managing within the capacity of individual watersheds.	Managing within environmental limits.
Government policies and direction not fully integrated.	Clear, government-wide policy, directions and outcomes on water.	Integrated outcomes on air, water, land and biodiversity defined in the place.
Traditional command and control regulatory system.	Much broader, innovative tools for watershed management.	Much broader, innovative tools + an aligned and enhanced regulatory system.
Desire by Albertans to be involved in their community.	Local, regional, and provincial partnerships established for planning and stewardship.	Place-based partnerships broadened to integrate air, water, land and biodiversity.
Pockets of alliances with stakeholders that achieve results.	Broad-based alliances with all parts of society to share responsibilities for outcomes.	Broad-based alliances with all parts of society to share responsibilities for integrated outcomes.
Meeting environmental standards.	Sustainability drives continuous improvement approaches.	Cumulative effects management drives innovation.
Focus on minimising and mitigating adverse effects.	Focus on improved quality of aquatic ecosystems and sustainability.	Focus on addressing environmental cumulative effects.

Source: Yee (2012).

representatives of state administration take crisis management in hand, in co-operation with these consultative bodies. This dual time-frame ensures that management is socially acceptable and fair in the long term, while preserving the rapid response times essential to crisis management.

River basin agencies (combined annual budget of EUR 1.9 billion) are entirely financed by water taxes and charges, according the "water pays for water" and "polluter pays" principles. But these taxes and charges mainly relate to industry and households. For example, differences in water pollution charges borne by farmers and households as well as additional water treatment costs due to diffuse agricultural pollution[17] sum up to between EUR 500 million and EUR 1 billion (Bommelaer and Devaux, 2011). Moreover, when redistributing the proceeds, there is a significant transfer from industry and households to agriculture. In other words, farmers contribute the agencies much less than they receive from them. The magnitude of this imbalance in taxation and public financial support shows that water users do not enjoy equal treatment.

In contrast to France, which relies on participatory democracy to manage tradeoffs in river basins through well-established river basin committees, a more "top-down" approach applies in **England and Wales**, with the Environment Agency overseeing implementation of the EU Water Framework Directive (WFD) in individual catchments and larger river basins. As part of its overall water risk management strategy, the Environment Agency for England and Wales delivers an integrated approach to water management and planning, based upon individual catchments and larger river basins. It has responsibility for managing and allocating water resources, protecting and improving water quality, conserving fisheries and aquatic biodiversity, and managing flood risk. It takes a strategic view, including assessing the impacts of climate change and population growth, which it uses to inform its operational activities at a local level. Planning for uncertainty is a major consideration, and it aims to ensure that management and investment decisions are resilient, adaptive and flexible to cope with the significant uncertainties and risks over coming decades.

The importance of water stewardship has been recognised: the Agency has moved from being only a regulator to an agency working with catchment stakeholders. The river basin scale is considered too big for water resource management in England and Wales, which is densely populated and has less natural water per person than many Mediterranean countries. Catchments are much smaller and are aggregated into river basin scale to satisfy the requirements of the EU Water Framework Directive (WFD) (2000).

A fundamental principle of water management in England and Wales is "no deterioration", in line with the requirements of the WFD. Rivers and wetlands of national and international significance are protected and improved by means of the most cost-effective measures. For non-designated rivers, action must be cost-beneficial.

Increasingly, an ecosystems services approach is being used to better assign values to the environment, and the benefits which healthy ecosystems can bring to drinking water, agriculture, flood risk management, recreation etc. Although paying for ecosystem services thinking is still in its infancy, it is starting to engender a different mindset among regulators, and users of water.

Charges on discharges (about GBP 60 [USD 108] million/year) cover much of the Agency's activities in water quality management, but some funding is required from general taxation. Pollution markets have not so far developed, although increasingly water companies are recognising that it is economically beneficial to work with farmers and land managers to control pollution at source.

Notes

1. OECD Expert Workshop on "Water Security: Managing Risks and Trade-offs in Selected River Basins" (Paris, 1 June 2012). The full text of case studies presented at the Expert Workshop can be accessed at *www.oecd.org/water*.

2. Case study prepared by Dominic Skinner (Water Science, Policy and Reform, The University of Melbourne).

3. Case study prepared by Beverly Yee (Alberta Environment and Sustainable Resource Development, Government of the Province of Alberta).

4. Case study prepared by Atika Ben Maïd (Department of the Commissioner-General for Sustainable Development, Ministry of Ecology, Sustainable Development and Energy, MEDDE/CGDD).

5. Case study prepared by Ian Barker (Water, Land and Biodiversity, Environment Agency for England and Wales).

6. Case study prepared by Eric G. Reichard, Zhen Li and Michael Land (*US Geological Survey*, California Water Science Center, San Diego, California) and Theodore Johnson (Water Replenishment District of Southern California, Lakewood, California).

7. When water and land were bundled together, salinity impact assessments became part of the trading processes, and it was only with the unbundling of water that this was no longer needed.

8. There were instances during the 2007-08 drought where environmental managers did have access to flexible arrangements. This was the case when the non-activation/reduced extraction of water licences made water available for use for the environment downstream (water shepherding).

9. Drought reduced irrigation water use to only 3 000 GL in 2007-08.

10. Proceeds from the abstraction charge cover all the Agency's activities in monitoring, assessing and managing water resources – some GBP 150 (USD 270) million/year.

11. Sharing of water quotas among farmers is subject to prefect approval.

12. Monthly flow that cannot be exceeded a given year with a probability of 1 on 5.

13. Contractual arrangements can be on the initiative of local authorities, river basin agencies and water companies (public water supply and bottled spring water).

14. Insurers have a statutory obligation to set aside part of the natural disaster premium (12% for a home insurance policy, 6% for a vehicle insurance policy) for the Barnier fund. Vehicle insurance is compulsory in France and 99% of the population have home insurance.

15. A Watermaster was established under the adjudications to act as the court-appointed administrator of the judgments and to record annual pumping amounts and enforce the terms of the adjudications.

16. The risk of flooding is mainly managed by central government and local authorities.

17. Including the costs to counter the proliferation of green algae, treat drinking water, mix raw water and transfer water from less polluted catchments.

References

2030 Water Resources Group (2030 WRG) (2009), *Charting Our Water Future: Economic Frameworks to Inform Decision Making*, *www.2030waterresourcesgroup.com/water_full/Charting_Our_Water_Future_Final.pdf*.

Agence de l'eau Seine Normandie (2009), *Protection des captages pour l'amélioration des pratiques agricoles*, Master's dissertation.

Andreen, W.L. (2011), "Water Law and the Search for Sustainability: A Comparative Analysis", in *Water Resources, Planning and Management:Challenges and Solutions*, R.Q. Grafton and K. Hussey (eds.), Cambridge University Press, Cambridge, England.

Bark, R. et al. (2011), "Water Values", in *Water, Science and Solutions for Australia*, CSIRO Publishing, Collingwood, Victoria, Australia.

Ben Maïd, A. (2012), "Water Security in France: Managing Risks and Trade-offs", case study prepared for the OECD Expert Workshop on "Water Security: Managing Risks and Trade-offs in Selected River Basins" (Paris, 1 June), *www.oecd.org/water*.

Bommelaer, O. and J. Devaux (2011), "Coût des principales pollutions agricoles de l'eau", *Études et Documents*, No. 52, General Commission for Sustainable Development (CGDD), Ministry of Ecology, Sustainable Development and Energy (MEDDE), September.

Briscoe, J. (2009), "Water Security: Why it Matters and What to Do About It?", *Innovations*, Vol. 4, No. 3.

Chiew, F.H.S. (2006), "Estimation of Rainfall Elasticity of Streamflow in Australia", *Hydrological Sciences Journal*, Vol. 51, No. 4, August.

Council of Australian Governments (COAG) (2004), *The National Water Initiative*, *www.nwc.gov.au/www/html/117-national-water-initiative.asp*.

Davis, C. and T. O'Rourke (2011), "Shakeout Scenario: Water System Impacts from a Mw7.8 San Andreas Earthquake", *Earthquake Spectra* 27 (2).

Department of Sustainability, Environment, Water, Population and Communities (DSEWPaC) (2012a), "Fact Sheet: Australian Government Investment in Enhanced Environmental Outcomes", DSEWPaC, Canberra.

DSEWPaC (2012b), "Fact Sheet: Environmental Flow Delivery Constraints", DSEWPaC, Canberra.

Fargher, W. and C. Olszak (2011), "Water Markets: A National Perspective", in *The Australian Water Project, Volume 1: Crisis and Opportunity: Lessons of Australian Water Reform*, J. Langford and J. Briscoe (Eds), Committee for Economic Development of Australia.

Garrick, D. and R. Hope (2012), "Economic Instruments to Manage Water Security Risks", *background Paper presented to the OECD Working Party on Biodiversity, Water and Ecosystems*, Paris 30-31 May, ENV/EPOC/WPBWE(2012)4.

Government of Western Australia (2011), *Groundwater Risk-based Allocation Planning Process*, Department of Water, Water Resource Allocation and Planning Series, Report No. 45, Perth, January, *www.water.wa.gov.au/PublicationStore/first/96735.pdf*.

Jones, L.M. et al. (2008), *The ShakeOut Scenario*, U.S.G.S. Open-File Report 2008-1150 and C.G.S Preliminary Report 25, U.S. Geological Survey and California Geological Survey, *http://pubs.usgs.gov/of/2008/1150/of2008-1150.pdf*.

Letrémy, C. and C. Peinturier (2010), "Le régime d'assurance des catastrophes naturelles en France métropolitaine entre 1995 et 2006", *Études et Documents*, No. 22, May.

Nicholls, R.J. et al. (2008), "Ranking Port Cities with High Exposure and Vulnerability to Climate Extremes: Exposure Estimates", *OECD Environment Working Papers*, No. 1, OECD Publishing, *http://dx.doi.org/10.1787/011766488208*.

OECD (2013), *Water and Climate Change Adaptation: Policies to Navigate Uncharted Waters*, OECD Studies on Water, OECD Publishing, *http://dx.doi.org/10.1787/9789264200449-en*.

OECD (2008), "An OECD Framework for Effective and Efficient Environmental Policies", *Paper presented at the Meeting of the Environment Policy Committee (EPOC) at Ministerial Level*, 28-29 April, *ENV/EPOC(2008)7/FINAL*.

Prosser, I. (2012), "Governance to Address Risks of Water Shortage, Excess and Pollution", unpublished *Paper presented at the OECD Expert Workshop* on "Water Security: Managing Risks and Trade-offs in Selected River Basins" (Paris, 1 June), CSIRO Water for a Healthy Country Flagship, Australia.

Reichard, E.G., Z. Li and C.M. Hermans (2010), "Emergency Use of Groundwater as Backup Supply: Quantifying Hydraulic Impacts and Economic Benefits", *Water Resources Research*, No. 46, W09524, *http://dx.doi.org/10.1029/2009WR008208*.

Rowan, R., N. Mitchell and J. Douglas (2006), "Water and Australia's Future Economic Growth", Industry, Environment and Defence Division, the Australian Treasury.

Skinner, D. (2012), "Australia: The Murray-Darling Basin", case study prepared for the OECD Expert Workshop on "Water Security: Managing Risks and Trade-offs in Selected River Basins" (Paris, 1 June), *www.oecd.org/water*.

World Bank (2004), *Water Resources Sector Strategy*, Strategic Directions for World Bank Engagement, World Bank, Washington, DC.

Yee, B. (2012), "Canada: Alberta's South Saskatchewan River Basin", case study prepared for the OECD Expert Workshop on "Water Security: Managing Risks and Trade-offs in Selected River Basins" (Paris, 1 June), *www.oecd.org/water*.

Chapter 3

Achieving water security targets through market-based instruments

Once set, water security targets should be achieved at least possible economic cost (i.e. cost-effectiveness should be pursued). This chapter suggests how market-based instruments can be used to promote more effective water management. Using theory, examples and case studies, a description is given as to how economic approaches may be used, particularly in OECD countries, to manage water risks.

In this Chapter, we suggest how market-based instruments can be used to promote more effective water management. The focus of this analysis is on specific aspects of water security that are amenable to the use of market-based instruments. Using theory, examples and case studies, a description is given as to how economic approaches may be used, particularly in OECD countries, to manage the risks of water insecurity.

There are a number of potential ways forward to improve water security management, such as supply management, demand management, quality improvement and the efficient and equitable allocation of water among uses. We investigate how market-based instruments may be used to improve how water risks are managed and, thus, improve water security. The focus is on four key issues: 1) water supply; 2) water demand; 3) water quantity; and 4) water quality, and how market-based instruments may help to achieve desired social objectives such as economic efficiency. We examine the importance of using market-based instruments as part of an integrated water resource management framework.

Principal findings include: 1) there are a variety of market-based instruments that can and should be used to provide efficient and effective management of water risks depending on the situation and context; and 2) market-based instruments should be used as part of a broader water management strategy which takes into account the needs of all stakeholders involved.

As water security issues differ across regions and countries, any discussion regarding the use of market-based instruments must be location-specific. While the focus is on the water security challenges facing OECD countries, it should be stressed that instruments which work effectively in nations with adequate financial resources for monitoring and enforcement, and a well-developed rule of law, may not be appropriate in countries which lack these attributes.

Economic principles and instruments

There are two key principles underlying the economic management of natural resources such as water – efficiency and equity.

Efficiency is concerned with maximising the welfare that is obtained from a resource by allocating it to its most valuable economic use. Resources such as water, for example, provide economic benefits in terms of human consumption, agricultural production, and services to the environment. In order to maximise the net benefits from the use of a resource, the marginal net benefit (the benefit from using an additional unit of the resource, less the cost of acquiring it) must be equalised across all uses, otherwise welfare could be increased by allocating more water to uses where the net benefits are greater. Such an allocation is said to be Pareto efficient; no one can be made better off without making somebody else worse off (Dasgupta and Heal, 1979).

Efficiency is a crucially important matter in water security because the welfare impacts of poorly allocated water resources can be very large, particularly for households with low incomes. Not only is it important in terms of ensuring affordable access to water, but also the large capital costs associated with water supply and delivery make it a fiscally important issue for many countries. For example, the combined cost of investments in water security in OECD countries amounts to tens of billions of dollars annually. Further, due to the lock-in effect of investments in large scale water infrastructure such as dams and desalination plants, the welfare impacts on humans and the wider environment are felt far into the future.

Equity is concerned with the fairness of the allocation of resources, or how they are distributed across a given population. Since efficiency only requires that total welfare is maximised, there is no requirement that the outcome is equitable. As a result, equity objectives may sometimes conflict with efficiency objectives (Howe, 1996). Raising the price of water for a particular use, for example, may increase the efficiency of water allocation, but the outcome may be less equitable if poorer households have to pay more for their water and are made substantially worse off. Thus, while efficiency is focused on maximising the benefits from water resources, equity is concerned with how to distribute the benefits in a non-discriminating manner.

Equity is crucially important in water security because it is widely recognised that a minimum supply of potable water is a vital prerequisite for life, health, dignity, and the realisation of other human rights.[1] In September 2010, the United Nations (UN) Human Rights Council affirmed by consensus that the right to water and sanitation is derived from the right to an adequate standard of living, recognising water as an inalienable human right.[2] UN projects such as the Millennium Development Goals (MDGs) recognise that, due to the benefits to human life, providing access to safe drinking water is a key goal.

However, there are many misconceptions regarding the right to water. It does not entitle everyone to an unlimited supply of water at all times, in any place, under any circumstances. The right to water does not obligate nations to share their water resources with other nations, as state sovereignty is unimpaired. The fact that water is a human right does not mean that it should be free, any more than health care is free. In the OECD, the issue of equity is much less pronounced because water is readily accessible for the vast majority of OECD residents. Nevertheless, providing water and sanitation services that is affordable to the poorest households remains an important policy objective in many countries. The MDGs strive to halve the proportion of the world's population without sustainable access to safe drinking water and basic sanitation by 2015.

The twin objectives of efficiency and equity may be achieved through the use of market-based instruments; policy tools which influence behaviour through their impact on market signals rather than explicit regulation. Examples of market-based instruments include access charges, pollution permits, tradable user's rights, among others. Their principal value is to provide a market signal as to the scarcity value of water among competing uses, including for environmental purposes (Table 3.1).

While the emphasis of market-based instruments is usually on achieving an efficient allocation of water resources, it is important to consider equity aspects at the same time. The OECD has developed a model which has been used by a number of countries to develop water and sanitation management strategies aimed at achieving both equity and efficiency goals. The aim is to strike a balance between sustainable financing of water and sanitation, and affordability. The realisation of this aim of sustainable cost recovery is based on tariffs, taxes, and transfers – of which market-based instruments play a key role (OECD, 2009). While market-based instruments can be used to promote equity objectives, they are

Table 3.1. **Summary of possible market-based instruments for water security management**

Water security issue	Recommended market-based instruments	Advantages of use
Water supply	Marginal social cost pricing, incorporating the scarcity value of water.	Signals the optimal time to invest in water infrastructure so that supply is augmented efficiently.
	International and regional water markets.	Allows trade of water from areas of surplus to increase the water supply in areas of scarcity.
Water demand	Regional water markets.	Allows trade of water from low to high value uses creating incentives to use water efficiently and reduce demand.
	Marginal social cost pricing, incorporating the scarcity value of water.	Reduces demand for water during periods of scarcity.
Water quantity	Buy-backs of water user's rights.	Secures water for environmental flows and offsets economic losses.
Water quality	Emission permit trading for point and nonpoint pollution.	Allows pollution to be reduced from the lowest cost sources.
	Emission taxes.	Creates ongoing incentive for all sources to reduce pollution.

Source: Grafton (2011).

typically focused on achieving efficiency. Nevertheless, equity goals can be addressed by imposing restrictions on how market-based instruments are applied or in terms of the initial allocation of user's rights. Given that equity considerations differ across countries, the focus of this section is on how market-based instruments can be used to provide efficient allocation of water resources.

Water supply

For many countries, freshwater resources depend on the level of surface run-off available. Since run-off can vary widely, the challenge of water supply lies in managing the natural variability of water resources to balance supply against demand. There are two main ways in which water supply can be managed: investing in large-scale water infrastructure and inter-basin water transfers.

Water infrastructure to deal with issues of water scarcity has historically been the focus of many countries' approach to managing water security. Large-scale water infrastructure can capture and store surface water run-off such as through dams, or augment natural freshwater resources, such as with desalination plants (Box 3.1).

The natural variability of water resources and the large upfront costs related to water infrastructure investment can make such projects highly risky for private investors. This business risk, together with the public good nature of ensuring an adequate water supply, often results in infrastructure investments made by the public rather than the private sector. The important public role in instigating and managing water infrastructure investments requires economic regulation.

The problem with adopting economic regulation to infrastructure investment is that water authorities and regulators are often not fully independent from governments, and this may lead to the politicisation of water supply decisions. Large-scale infrastructure projects, in particular, may be chosen by governments on grounds other than those that promote the most efficient method of delivering water (Sibly and Tooth, 2008). Investments in water infrastructure can also have significant long-term impacts on welfare, in terms of costs to the taxpayer and environmental impacts. In order to reduce these costs and maximise welfare, a comprehensive cost-benefit analysis, that includes all economic,

Box 3.1. **Water infrastructure investment options for Adelaide, Australia**

In 1995, Adelaide had the lowest ratio of water storage capacity to annual water demand of all Australian capitals. As the population of the city has continued to grow and surface water run-off has decreased, the amount of water storage relative to demand has fallen further. In order to meet the predicted shortfall in water supply, it is estimated that the water supply needs to be increased by an additional 164 GigaLitres (GL, or 164 billion Litres) per year, and a number of different options have been proposed to achieve this target.

Desalination – Desalination plants, where salt or brackish water is converted to freshwater by the removal of salts and minerals, can be used to provide large, reliable supplies of good-quality water since they are independent of rainfall and do not reduce the amount of water available for other uses. A desalination plant has been proposed in Adelaide which would produce 100 GL per year at a cost of AUD 2.30 (USD 2.40) per kiloLitre (kL, or thousand Litres). In order to limit the environmental impacts, its proponents have specified that renewable energy would be used in its construction and studies would be undertaken to protect the surrounding ecosystems.

Recycling urban stormwater – A second proposal is to capture urban stormwater run-off and store it in underground aquifers. Although dependent on rainfall, stormwater is not as variable as run-off from rural catchments, and the level of run-off is increasing due to urban expansion and increasing amounts of impervious surfaces. Further, an increasing intensity and frequency of storms predicted to occur with climate change may increase the amount of stormwater that can be captured. In Adelaide, it is proposed that recycling urban stormwater could provide up to 60 GL per year by 2050 at a cost of AUD 1.20-2.00 (USD 1.30-2.10) per kL for non-potable water and AUD 1.30-1.70 (USD 1.4-1.8) per kL for potable water.

Recycling treated sewage water – A third proposal is to recycle treated sewage water instead of discharging it into river systems or the sea. As this water is slightly saline it can be diluted with stormwater and used for non-potable water supplies, such as toilet flushing and watering gardens or, if treated further, used as drinking water. In Adelaide, recycling around 33% of treated sewage water could increase water supplies by 50 GL per year at a cost of AUD 1.90 (USD 2.00) per kL of non-potable water and AUD 2.50 (USD 2.60) per kL of potable water.

Rain water tanks – Installing rainwater tanks increases the amount of water that is captured usefully and enables households to harvest water directly for drinking, gardening, cooking and washing, etc., although it also affects the time and size of groundwater recharge and downstream surface flows. In 2007, 38% of households in Adelaide had rainwater tanks installed and, since July 2006 it is required that all new homes have rainwater tanks installed. Subsidies for installing rainwater tanks are also available for existing homes. Such measures are estimated to increase the water supply by 5 GL per year at a cost of AUD 3.75 (USD 3.90) per kL.

Expanding reservoir capacity – A final proposal is to increase the capacity to store water in reservoirs or dams. Reservoirs in the Mount Lofty Ranges supply most of Adelaide's water and expanding their capacity could increase the water supply by 40 GL at a cost in excess of AUD 2.40 (USD 2.50) per kL, although this expansion risks reducing already stressed environmental flows and destroying the habitat of endangered species.

Source: Dillon (2011).

social, and environmental costs, should be carried out for each potential investment (Becker, Lavee, and Katz, 2010; Birol, Koundouri, and Kountouris, 2010; Fisher et al., 1995; Wittwer, 2010).

A potential objection to the use of cost-benefit analysis to determine public investment in water infrastructure is that access to safe water is a human right. According to this view, investment in water infrastructure should be determined by government regulation rather than the estimated returns on investment. However, basic human rights exist in many forms including rights to shelter, basic health care and education that also require substantial government investment. Thus, money spent on securing one human right can mean less money is spent securing other rights. While access to safe water is extremely important, it does not mean that all government investments in water infrastructure are appropriate. Indeed, whether investment in water infrastructure is justified does not solely depend on the law or human rights *per se*, but rather on the willingness and, crucially, the ability of governments to make such investments. In order to ensure that government resources are used to achieve the highest social returns, investment decisions should be made using cost-benefit analysis and other tools to assess the trade-offs between increasing water access and the costs of providing access. While evidence suggests that the rates of return for investment in water infrastructure are typically higher than other infrastructure or social investments, this does not mean that it will always be the case.

Good-quality cost-benefit analysis requires accurate estimations of the future benefits and costs of an investment. While the costs of proposed investments are often predictable, the natural variability of water resources makes predicting the benefits less straightforward. During a prolonged period of low rainfall, for instance, a large investment in water infrastructure may appear to be beneficial. However, water levels in catchments may subsequently increase and the additional supply augmentation may no longer be required.

If water prices are set efficiently then the optimal timing for investment is signalled by market prices: a fall in water availability pushes up water prices and makes infrastructure investments profitable; thereby increasing water supply and balancing the supply and demand for water. However, water authorities often set prices without proper consideration of efficiency. For instance, for residential water pricing, regulators will typically claim to set volumetric prices equal to long-run marginal (private) costs of supply or the average levelised cost of the next most affordable water supply, rather than the scarcity value of the resource (Grafton and Kompas, 2007; Sibly, 2006a; Sibly and Tooth, 2008). As a result, prices do not fully reflect the scarcity of the resource which weakens the ability of prices to signal when to invest in water infrastructure, and so can lead to significant welfare losses (Box 3.2).

In order to deal with the limitations of the economic regulation approach to investment in water infrastructure, three alternatives approaches have been proposed using market-based instruments. *First*, Sibly and Tooth (2008) propose the separation of water infrastructure from water storage (unbundling) by introducing competition within water supply. They argue that the introduction of competition within water supply will lead to more efficient water pricing and signal optimal investments in water supply augmentation, as well as providing incentives for private investment in infrastructure. The view that vertical separation of bulk water supply from retail distribution and also horizontal separation of the retail and waste-water sectors can generate efficiency gains is currently being explored in Australia's urban water sector (Productivity Commission, 2011).

Second, an alternative approach is to offer a portfolio of water contracts to consumers. Households who want greater water security and who wish to avoid any mandatory water restrictions would opt for higher priced but high security water. By contrast, households

> ### Box 3.2. **Investing in desalination in Sydney, Australia**
>
> In 2007, a contract for a desalination plant was signed in Sydney due to concerns over water shortages. However, the construction of the plant took several years, during which the ending of the drought alleviated some of the water security concerns. Following the construction of the plant, water prices increased by 50% from 2007 to 2010 to cover the costs of investment. By contrast, if scarcity prices had been introduced in Sydney prior to building the desalination plant, the market would have sent signals about the optimal time to invest in desalination. By estimating the optimal time to invest in desalination based on efficient volumetric prices, Grafton and Ward (2010) found that the investment in desalination in Sydney was made prematurely, and led to welfare losses valued at hundreds of millions of US dollars per year. These losses arose from the costs associated with using mandatory water restrictions rather than dynamically efficient pricing and, ultimately high volumetric water prices needed to cover the high capital costs associated with the premature construction of the desalination plant.
>
> *Source:* Grafton and Ward (2010).

less concerned with reliability may choose a lower water price contract, but which also includes a greater likelihood of restrictions (Productivity Commission, 2011).

Third, a number of studies suggest that water should be priced volumetrically based on the supply of water available so that price reflects the scarcity value of the resource (Grafton and Kompas, 2007; Grafton and Ward, 2010; Hughes, Hafi, and Goesch, 2009; OECD 2008; Sibly, 2006a; Zetland, 2011). Under this approach, the price of water would signal the optimal time to invest in water supply augmentation as well as raising revenues needed to cover the investment (this issue of scarcity pricing will be looked at in more detail in the section on water demand).

While the focus of this section has so far been on investing in water infrastructure to deal with issues of water scarcity, an equally important role of **water infrastructure is dealing with periods of excessive water supply**. While flooding can never be completely prevented, the economic, environmental, and social impacts of floods can be significantly reduced through investment in water infrastructure such as dams, levies, flood defences, drainage systems, and stormwater management systems.

Like water infrastructure designed to deal with periods of water scarcity, the public good nature of investments in flood management mean that it is under provided in private markets, thereby leaving a significant role for government (Shaw, 2005). However, investments in infrastructure to deal with excess water supply often overlap with investments which deal with water scarcity and this can create competing objectives for decision makers involved with flood management: from avoiding loss of agricultural output, property damage, and loss of life; to retaining water for household, irrigators, and hydropower; as well as securing the provision of flows for the environment. These often conflicting objectives mean that effective flood management requires a balancing of the trade-offs between flood prevention, storing water for consumptive purposes, and providing sufficient environmental flows (Box 3.3).

In order to deal with these trade-offs in a clear and transparent manner, a full cost-benefit analysis should be carried out for each infrastructure investment decision. The use of cost-benefit analysis is particularly important for large scale infrastructure which has

> **Box 3.3. Flood management and the Wivenhoe Dam in Brisbane, Australia**
>
> Over the last 40 years, the city of Brisbane has experienced significant problems with both drought and flooding. After severe floods in 1974, the Wivenhoe dam was built to reduce the impacts of future floods and act as a store of water during times of scarcity. In order to achieve the joint goals of water storage and flood prevention, the dam was designed to meet the region's drinking water supply with an additional 125% excess capacity to cope with flooding. The design of the dam creates risk-risk trade-offs between the goals such that when more water is stored in the dam, the chance of water scarcity problems is lower, however, there is less capacity to capture flood water so the dam is less effective in terms of flood prevention.
>
> In 2008, during a period of prolonged drought, the water level fell to around 17% capacity leading to a focus on managing water scarcity. However, after several months of intense rains in 2010 the water level increased at a rate of 1 million MegaLitres (or 1 trillion Litres) per day leading to significant flooding throughout the city and surrounding area. While the dam likely reduced the impact of the floods compared to not having a dam, the operational rules of the management of the dam resulted in water being kept in the dam when it could have been released earlier. Earlier release would, in turn, have reduced the impact of the flooding that actually occurred and that resulted in property damages worth about half a billion Australian or US dollars.
>
> The experience of Brisbane highlights the complex trade-offs that are present in flood management schemes. On one hand, the floods may have been much more significant without the Wivenhoe dam which was designed to reduce the height of a severe flood peaks by up to 2 metres. On the other hand the lengthy period of water scarcity prior to the flood contributed to a reluctance to release water when the dam was close to being full.
>
> *Sources:* Barry (2011); Pittock (2011).

significant impacts such as the construction of large dams. In most OECD countries, the rate of dam building has fallen significantly because most of the sites where the benefits to dam construction exceed the costs have been exploited (WCD, 2000). Further, there is growing recognition of the social and environmental impacts of dam construction and these impacts are increasingly being incorporated into cost-benefit analysis through the use of non-market valuation techniques, thereby making it more likely that costs of dam construction exceed the benefits (Shaw 2005). In the past, dams have often been constructed without adequate cost-benefit analysis and now that the importance of environmental impacts are recognised there is increasing debate over whether to remove such dams in order to allow more natural environmental flows (Shaw 2005).

Due to the complex nature of the trade-offs involved in flood management, the relative benefits and costs of flood infrastructure are often controversial. One method of resolving such controversy and building consensus is to develop an integrated flood and drought management strategy which encourages the participation of multiple stakeholders. An example of this participatory approach to water supply management is the Upper Iskar Basin in Bulgaria, where a wide range of regional stakeholders including government ministers, private companies, NGOs, local council members, national experts, and local residents, were involved in the design of an integrated flood and drought management strategy (Daniell, Ribarova and Ferrand, 2011). This issue of integrated water resource management is discussed further below.

Further, in addition to the increasing variability in water supply due to climate change, growing rates of urbanisation are increasing the damages from flooding due to higher population densities on flood plains and greater levels of stormwater runoff due to the increase in impermeable surfaces. The risks of flood plain development suggest that strict planning regulations should be in place and, from an economic perspective, it is critically important to ensure that people face the actual flooding risks and are given the best available information about these risks so that they can adapt their behaviour. Where people do not confront the actual flooding risks or are unaware of them, for example if the government provides full compensation for flood damages or do not provide maps of flood prone areas, it generates perverse incentives and can encourage development in flood zones that would otherwise not occur.

An alternative method of managing water supply is to physically transfer water from one basin to another. Typically, **inter-basin water transfers** involve regional transfers across basins via aqueducts pipes and through existing water channels.

Due to variability in rainfall, water availability is not homogeneous across countries or between regions within countries. For instance, some areas may have excess water at the same time that others are experiencing water scarcity. This gives rise to different marginal values of water across different areas; water, for example, is less valuable in a flooded area than in a drought affected area. This difference in the marginal value of water creates gains from trade if water resources can be spatially reallocated to their most valued use. The standard economic approach to resolving differences in the marginal values of resources is through the trade of water.

Currently, international water markets are poorly developed and human-directed international water transfers are uncommon; usually being adopted as a short-term measure to increase water supply in emergency situations.[3] In 2008, for example, prolonged periods of drought in Catalonia led to Barcelona importing tankers of freshwater from Tarragona and Marseille in order to alleviate the impacts of the drought (Burnett, 2008).

International water transfers can also be made over longer periods, with ongoing supplies of water provided over a number of years. This, however, requires significant investments in large-scale water infrastructure, such as pipelines and aqueducts. As well as involving significant costs in terms of construction and maintenance, such investments can also have significant social and environmental impacts if large amounts of water are removed from an area over long periods of time. They also involve difficult political negotiations; as demonstrated by Barcelona's failed proposal to import water from the Rhone river via a 320 km aqueduct from Montpellier, France (Gleick et al., 2002). Likewise, the "Peace Water Pipeline", designed to pipe water from Turkey to Syria, Jordan, and the Arab Gulf States, is another example of a water transfer project that could not be realised due to the reservations of water exporting countries (Gruen, 2007).

One of the reasons why international water markets are poorly developed is that the costs of purchasing and transporting large volumes of water are often very high. In Barcelona, for example, the average cost of imported water was around USD 8 per kL (Burnett, 2008), compared to around USD 2.00 per kL for water produced from desalination plants. Furthermore, the legal status of international water trading is not yet clear as it does not fall under the World Trade Organisation (WTO) agreements and currently there is no international legislation governing cross border water issues.[4] Concerns have also been raised over the equity of treating water as a tradable, economic good; with the worry that trading water, which is often seen as a basic human right, not a commodity, may lead to overexploitation of the resource (Gleick et al., 2002).

Most inter-basin water transfers are, therefore, done within countries on a regional basis, particularly in large countries with significant internal heterogeneity of water resources such as Australia, Canada, China, India and the United States. Although these transfers are within national borders, they are often controversial and generate considerable public concerns due to their potentially large environmental costs (reduced environmental flows, loss of natural habitats, etc.) and social costs (lack of access to water for indigenous people, relocation of communities, etc.). Despite the potential magnitude of these costs, there have been few cost-benefit analyses of inter-basin trading, and the full impact of such schemes, particularly in terms of their environmental impact, have not fully been explored (Ghassemi and White, 2007).

As water scarcity problems increase, inter-basin water transfers are a potentially useful method of alleviating water supply problems in cases of serious drought. However, due to the political difficulties and high costs of international transfers, they are likely to predominantly remain a regional solution within large countries with heterogeneous resources. While it has been argued that globally integrated water markets are likely to develop within the next 25 to 30 years once spot markets for water are integrated (Business Insider, 2011), it is more likely that international water trade will remain insignificant due to the high costs of transporting water. While future markets for water will develop, they are likely to be limited to locations where water can be delivered at a reasonable cost, i.e. within national borders and within specific river basins.

In areas where water scarcity is particularly acute, there may be greater scope for international water transfers. In the Middle East, for instance, despite significant political differences, water transfers are considered a potentially viable option and it is argued that a market-based approach may reduce the risk of future conflict over water resources, as it may reduce pressure on scarce water resources which are accessed by multiple countries (Gruen, 2007; Rende, 2007; Wachtel, 2007).

Water demand

There are two principal ways in which water demand can be managed using market-based instruments: firstly by establishing water markets (which will be looked at in an agricultural context); and secondly by setting efficient water prices (which will be looked at in an urban context).

The problem of over-extraction of water resources often reflects a lack of clearly defined legal rights of water users. Typically, an individual using a water resource cannot prevent others from withdrawing water from that resource, and so has an incentive to take the water they need before others do so. This can result in the overuse of water resources. Furthermore, withdrawing water from a resource impacts the ability of others to do so; upstream diversions, for example, reduce the flow of a river downstream. The presence of these negative externalities mean that individuals do not take into account the full cost of their decisions when extracting water and may only take into account the personal cost of withdrawing water, not the cost this will have on others. This creates a wedge between the marginal cost of water withdrawals faced by individuals and the marginal cost of water withdrawals faced by society, and this divergence can lead to over-extraction of the resource.

The standard economic solution to such problems is to put a cap on the amount of water that can be extracted by assigning a fixed number of tradable user's rights for accessing the resource. In an efficient market, individuals face the full cost of using the resource and the user's rights can be traded to those who value water the most, thereby ensuring that the resource is allocated efficiently. Thus, in order to reduce demand for water and ensure that it is used efficiently, **agricultural water markets** can be created which put a limit on extractions and allow the buying and selling of water user's rights (Easter, Rosegrant and Dinar, 1998; Howe, Schurmeier and Douglas Shaw Jr., 1986; Milliman, 1959; Saliba and Bush, 1987).

Under a properly functioning market (where the number of user's rights is not over-allocated), access to the resource is no longer freely available, but depends on acquiring the water user's right. This creates a scarcity value for water and an incentive for individuals to use water more efficiently, for example, by employing water-efficient technology, adopting deficit irrigation, or growing less water-intensive crops; thereby reducing demand. Further, markets allow water to be transferred from low-value uses to high-value uses. Urban water prices, for example, are often much higher than the price of agricultural water, even after accounting for the differences in quality (Grafton et al., 2012). Therefore, allowing some of this water to be traded from lower-valued agricultural use to higher-valued urban use can be beneficial to both buyers and sellers.

In response to rising water demand and increasing water scarcity, the use of water markets to manage water demand is increasing, and well-developed markets exist in Australia, Chile, and the United States. The type of water markets differs widely across countries, depending on the historical context of water law and the priorities of the government when setting up markets (Box 3.4).

Box 3.4. **Water markets in the Western United States, Australia, and Chile**

Western United States

In Western United States, most legal rights of water users are through prior appropriation; where the first party to use a water resource holds the right to continue using it in this manner until they decide to sell or lease the user's right. These water user's rights are conditional in that the water is owned by each state and individuals are permitted to use them by each state subject to certain conditions so that the resource is managed in the public interest. These user's rights can be capitalised into the value of land and traded, or leased and sold separately.

Prior appropriation allows use of a fixed quantity or flow of water for diversion from a stream or withdrawal from an aquifer, and they are divided into junior and senior user's rights. Senior right holders receive their water first followed by junior right holders. Thus, in times of drought, junior right holders may lose some or all of their water access and are dependent upon return flows from senior right holders. The trade of appropriative water user's rights in the United States may therefore impair third parties and so is subject to state regulation to ensure that "no harm" is inflicted on junior right holders.

Appropriative user's rights are also conditional upon water being used beneficially and, if use is not judged to be beneficial, the right may lapse. Initially, irrigation was the dominant basis for defining beneficial use; however, this has recently been expanded in many jurisdictions to include environmental flows. The legal intent of the principle of denying water user's rights for water that is not used beneficially is to remove the incentive

Box 3.4. **Water markets in the Western United States, Australia, and Chile** *(cont.)*

for wasteful use. In practice, however, the beneficial use condition can lead to inefficient water allocation as right holders withdraw water even if the marginal value of use is low, in order to prevent the right from being lost.

Murray-Darling Basin, Australia

Unlike appropriative user's rights which evolved in Western United States, Australian surface water user's rights began as riparian rights; where landowners with property adjoining a water resource were granted user's rights. Over time these were transformed to statutory water user's rights. The over-allocation of statutory user'srights led to increasing pressure for legal rights of surface water users to be separated from land so they could be traded. As a result of this pressure, reforms in 1994 and 2004 led to statutory water user's rights being fully separated from land rights, surface water extractions from the Basin being capped at 1994 levels, and a fully established surface water market.

Statutory water user's rights in the Murray-Darling Basin provide an entitlement to a specified share of surface water from a consumptive pool. However, the amount of water that entitlement holders are allowed to withdraw depends on the seasonal allocation of water for that entitlement. Seasonal water allocations are based on the amount of water in storages and the expected inflows. The reliability of water user's rights varies from high reliability where, in most years, holders of these rights receive their full allocation, and low reliability where, in dry years, there may be lower or even zero allocations of water. Both water entitlements and seasonal allocations are available for trade.

This system reduces the third party impacts of water trade so there is no need for a "no harm" condition and, combined with the absence of a "beneficial use" condition, Australian water markets are generally more efficient than those of Western United States. There are, however, restrictions on trade, particularly on exporting water out of basins, and as water entitlements are statutory user's rights, they can be repealed or changed by the government without compensation, although this is unusual in practice.

Chile

Water markets in Chile evolved from a complicated history of water law that first developed at a local level with ambiguous definitions of water user's rights. Agrarian Reform Law in 1951 weakened private user's rights and increased the role of the state. In 1981, large-scale reform defined ground and surface water as national property available for public use through the granting of tradable user's rights of water to private users.

The 1981 reforms granted existing water user's rights for surface and groundwater without cost. New or unallocated water user's rights were auctioned to users and then sold or leased. The market was similar to Australia in that both permanent and contingent user's rights could be traded, where contingent rights provide allocations only when there is sufficient availability. There was also trade of permanent water user's rights and seasonal allocations as in Australia.

As a result, following the 1981 reforms, water user's rights in Chile were separate from land, freely tradable, subject to minimal regulation, and governed by civil rather than administrative law. However, because the 1981 reforms did not address issues of multiple water uses, conflicts have been resolved through negotiation and bargaining among users, with the courts having the final say in case of an unresolved disagreement. This lack of clarity over multiple uses led to confusion over the priority of consumptive and non-consumptive user's rights.

> **Box 3.4. Water markets in the Western United States, Australia, and Chile** (*cont.*)
>
> In 2005, a reform of the 1981 Water code was approved to address social equity and environmental protection issues. Among the most important aspects of the reform are the following: 1) granting authority to the President to exclude water resources from the market when necessary to protect the public interest; 2) the obligation of the General Water Directorate to consider environmental aspects in the process of establishing new water user's rights, especially when identifying ecological water flows and protecting sustainable aquifer management; and 3) charging permit fees for unused water user's rights and limiting applications for water user's rights to the original needs in order to prevent hoarding and speculation. In addition, both the Ministry of Environment and the General Water Directorate of the Ministry of Public Works are currently working on the regulation that establishes the minimum ecological flows.
>
> *Source:* Bauer (1997); Grafton et al. (2011a).

Despite the increasing use of water markets, and the increasing severity of water scarcity problems, they remain an underused tool in managing water security. The reluctance to use water markets to manage water demand in many places is due to concerns over a number of issues.

One of the principal objections is that water trading can have negative impacts on third parties which are not represented in the cost of the trade. At an individual level, purchasing a water allocation upstream may reduce downstream flows in times of drought, for example. However, if water user's rights are granted as shares of the annual water availability and so are adjustable based on levels of run-off, as in Australia, then junior parties would not be differentially impacted and potential third-party harm from trades would be reduced (Grafton et al., 2012).

At a regional level, there are also concerns that there may be significant third-party impacts on communities which are dependent on water use intensive industries. While such concerns are legitimate, the evidence suggests that the impacts are not as significant as is often feared. An evaluation of water markets in California concluded that although there were some negative effects locally, overall water transfers increased total welfare (Howitt, 1994). Further studies suggest that issues other than regional water trade have much bigger (positive and negative) impacts on communities than water trade (National Water Commission, 2010).

While potential third-party impacts appear to be smaller than is often claimed, some individual towns and communities may be made worse off. These impacts could be reduced through compensation to local communities to enable them to transition away from water-intensive uses (Haddad, 2000). Alternatively, caps on the volume of water which can be traded could be implemented to further reduce third-party impacts, or state governments could adjust local government funding to account for water trade effects (Chong and Sunding, 2006).

Another objection to water markets is that, under standard economic theory, trading only leads to efficient outcomes when there are no transaction costs (Coase, 1960). In reality, transaction costs for water trading may be significant: trading may require moving large quantities of water, regulation may put restrictions on certain trades, there may be no single place for trade, making it difficult to arrange trades, negotiating water trades can take a significant period of time, and the impacts on third parties can create legal costs as those who are harmed seek compensation. Such costs may prevent water markets from working effectively, as the gains to trade may be outweighed by the costs.

Evidence from a number of studies suggests that, although transaction costs for water trades can be significant, this does not mean that water markets are ineffective (Carey, Sunding and Zilberman, 2002; Grafton et al., 2012; Grafton et al., 2011a; Hearne and Easter, 1997). Further, institutional design can increase efficiency by: reducing or eliminating exit fees, setting up trade exchanges, and reducing bans or limits on trade, etc. Colby (1995) also contends that transaction costs are not just the result of the inefficient consequences of regulation, but can be designed to take into account the externalities that water trades have on other users. Thus, transaction costs can be used as a tool to provide traders with an incentive to account for the social costs of water transfers.

A similar objection against water markets is that the conditions needed for competitive markets may not exist; markets may be too thin (too few buyers and sellers to be competitive), or dominated by large traders who manipulate prices or hoard water user's rights. This is a concern, and in some areas where there are few users competing for access to water resources, water markets may be inappropriate. However, in areas where there are multiple users and over-extraction is a serious problem, water markets are worthy of consideration.

A further objection is that water markets give too much weight to considerations of efficiency over equity (Bauer, 1998). While this is a legitimate concern, the criticism applies not to the theory of water markets in general, but to their particular design. As demonstrated by the developing water markets in South Africa where equity is given more importance than efficiency (Muller, 2009), this can be addressed in the institutional design of markets.

Thus, the potential challenges associated with water markets do not prevent the possibility of a well-functioning and socially beneficial markets (Chong and Sunding, 2006). Indeed, several studies suggest that water markets can both decrease water demand and increase overall welfare; provided they are well designed. A study of the impact of water markets in the Murray-Darling Basin estimates the annual gains from trade are around AUD 700 (USD 730) million in years of below normal inflows and AUD 300 (USD 315) million in years with above normal inflows (Peterson et al., 2004). The National Water Commission (2010) also found that water trading increased Australia's Gross Domestic Product (GDP) by AUD 220 (USD 150) million in 2008-09, and Gross Regional Product (GRP) of the southern Murray-Darling Basin by AUD 370 (USD 250) million, as well as having a positive impact on the environment by reducing overall water demand. Further, although drought in the Murray-Darling Basin led to a 70% decrease in irrigated water use from 2000-01 to 2007-08, the nominal gross value of irrigated agriculture fell by less than 1%. This is because farmers, by changing their practices and taking the opportunity to reallocate water to higher-value uses through water markets, were able to maintain their value of production (Grafton et al., 2012). Evidence from studies of water markets in other countries also suggests that water markets have had a positive impact in both Chile (Hadjigeorgalis and Lillywhite 2004; Hearne and Easter, 1997), and the United States (Grafton et al., 2012; Grafton et al., 2011a).

In summary, well-designed water markets can create incentives to use water efficiently which help to improve water use efficiency and allow farmers to cope with reduced water availability, and therefore have the potential to decrease water demand and increase overall welfare. Under schemes in which cap and trade are well designed (i.e. the cap is set locally and takes into account all externalities) water markets can promote efficient allocation of water and additional measures such as water abstraction charges will not be required. However, in some cases water markets are designed inefficiently and the cap on water extractions is determined by historical or political reasons rather than by full consideration of externalities and trade-offs. In such cases, the use of water markets combined with a water abstraction charge may improve efficiency. The criteria for ensuring that a water market is well designed depend on the local conditions. While there are broad criteria that should be met in all markets, such as not over-allocating user's rights, the particular criteria depend on the particular regional challenges and as a result the nature of water markets differs widely across countries (as illustrated in Box 3.4). It is also important to note that water markets are not always an appropriate solution; thin markets with high transaction costs and concentrated market power are unlikely to lead to an efficient allocation of water resources. Thus, we stress again that whether or not water markets should be implemented, and how they should be designed if they are implemented, depends on the local conditions and challenges faced.

One of the most efficient ways to regulate demand for a resource or good is through prices; as the price of a good is increased, consumers use less of it and demand falls.[5] In this section we focus on **urban water pricing**, although similar principles can be applied to agricultural and industrial water use.

There are two objections commonly made against the use of water pricing to regulate demand: *first*, that water demand is price inelastic, so that raising its price will not result in substantial reductions in consumption; and *second*, that higher water prices are inequitable because the proportion of household income spent on water increases for lower incomes and, thus, unfairly burdens low-income households.

In response to these concerns, many water utilities regulate water consumption, especially in periods of droughts, by using non-price instruments such as mandatory water restrictions, where certain types of water use are restricted at certain times, and measures that encourage water saving behaviours through the subsidisation of water efficient devices, education campaigns, and technological standards. While mandatory water restrictions can be effective in reducing demand, they can generate significant overall welfare losses relative to efficient volumetric water pricing (Garcia-Valinas, 2006; Grafton and Ward, 2008; Mansur and Olmstead, 2007; Woo, 1994). This is because water restrictions ignore heterogeneity in the marginal value of water across consumers and across uses (Sibly, 2006a). Further, the reduction in consumer surplus from water restrictions is lost to the economy whereas under price rationing, the loss in surplus is captured by government revenues (Allen Consulting Group, 2007).

With regards to alternative measures aimed at promoting water efficiency, the time lag required to install new technologies limits their ability to reduce short-run demand, and the size of the reduction needed in water consumption may be of a magnitude that imbalances in water supply and demand may still remain (Grafton et al., 2011b). However, the effectiveness of water demand management policies that include campaigns to promote water saving behaviours and water saving devices may be enhanced when combined with volumetric pricing (Grafton et al., 2011b).

Using data from 1 600 households across ten OECD countries, Grafton et al. (2011b) find that price-based approaches are likely to be the most effective and most efficient method of controlling long-run urban water demand. In their study they found that the price elasticity of demand ranged from -0.33 in Norway to -0.88 in Italy, with an average of -0.56,[6] and that households facing volumetric water pricing consume around 20% less water than those facing tariffs which are not directly linked to the volumes of water used.

Claims that water demand is unresponsive to price changes are further undermined by a number of studies which show that: 1) informing consumers about the volumetric price they pay on their water bill increases price elasticity (Gaudin, 2006); 2) consumers are more responsive to price changes the longer they have to adapt, so price elasticities are higher in the long-run (Renwick and Green, 2000; Dalhusien, de Grroot and Nijkamp, 2000; Worthington and Hoffman, 2008); and 3) price elasticity increases with higher prices, because at higher prices, water charges account for a larger share of household expenditures (Grafton et al., 2011b). Since many of the studies finding that water demand is unresponsive to price have looked at short-run, subsidised, water prices, the effectiveness of using price as demand tool may have been underestimated (Box 3.5). Indeed, the demand for water in the long-run appears to be responsive to changes in price, with general consumption per connection declining by around 1% per year for the last 15 to 20 years as water tariffs have increased.

Box 3.5. **Agricultural water pricing in Israel**

Due to increasing water scarcity, water prices in the agricultural sector in Israel rose 100% over the past decade. This increase in prices led to substantial changes in agricultural practices including: a move to drip irrigation; adopting more appropriate crops; removing water-intensive trees and replanting with water-saving types; and increasing the use of recycled and desalinated water sources. As a result, agricultural water demand has declined significantly and desalinated and recycled sources of water now make up around 50% of irrigated water use. Despite the significant decline in agricultural water use, efficiency gains have meant that agricultural production has actually increased. Further, higher water prices and increased use of alternative sources of water have stimulated technological innovation and exports of water technology grew by 21% in 2006 and 28% in 2007.

Source: OECD (2010a).

Grafton et al. (2011b) found that low-income households spend more than twice as much of their income on water bills than high-income households. Further studies looking at the issue in the OECD area confirm that the affordability of water in low-income households is a significant issue (OECD, 2009; OECD, 2010b). Thus, considerations of the equity of raising water prices to control demand are important. In order to reduce the burden of higher water prices on low-income households, a number of measures could be implemented, such as: reduced water access fees, progressive tariffs, water vouchers, or lump sum transfers. Further, if water pricing is designed well, increasing prices can actually improve equity, as the revenues can be used to increase water access among low-income households (Rogers, de Silva and Bhatia, 2002).

Water pricing frameworks differ considerably within the OECD area and common variants include: 1) flat rates, where consumers pay a flat rate for water regardless of their consumption (the rate can be uniform or differentiated by consumer type); 2) volumetric rates, where consumers pay a single rate per cubic metre consumed, these are often combined with a fixed access charge which can be uniform or differentiated; 3) increasing block tariffs (IBTs), where the volumetric rate increases with the amount consumed (blocks can be applied uniformly or differentially); or 4) decreasing block tariffs, where the volumetric rate decreases with the amount consumed (OECD 2010b).

Reviews of water pricing in the OECD area have shown that the use of flat fee systems and decreasing block tariffs has declined in favour of IBTs and volumetric rates with a fixed charge (OECD, 1999; OECD, 2003; OECD, 2010b). Volumetric rates have been adopted as they provide a stabilised source of revenue for suppliers and the variable component can be set to cover long-run marginal cost whilst the access fee can be reduced or eliminated for low-income households. Likewise, IBTs can theoretically be designed to allow free or low-cost water to low-income households who consume less water, and long-run marginal costs can be covered by higher tariffs for those who consume more (Rogers, de Silva and Bhatia, 2002). In the latest review of 184 water utilities across the OECD area: no flat fees were found; 90 used single volumetric tariffs (60 also included a fixed charge); 87 used IBTs (2 also applied a fixed charge); and 7 used decreasing block tariffs (all of them in the United States)[7] (OECD, 2010b).

There have been continued real water price increases over recent years, together with increasing separation of wastewater from drinking water charges, and most OECD countries are now moving towards water pricing based on meeting the costs of supply (including abstraction, treatment, and pollution costs), rather than subsidising access on equity grounds (OECD, 1999; OECD, 2003; OECD, 2010b) (Box 3.6).

Box 3.6. **Full-cost water pricing in Denmark**

Since 1992, urban water prices in Denmark have been based on full-cost recovery so that prices cover both economic (through user charges) and environmental costs (through taxes). All urban water users are metered and water prices are charged according to the volume consumed. Since the policy's introduction, water prices have risen substantially; during the period 1993-2004, the real price of water (including environmental taxes) increased by 54% and prices are now among the highest in the OECD area. The rise in prices has led to a substantial decrease in urban water demand from 155 to 125 Litres per person per day, one of the lowest levels in the OECD. Since water pricing is purely volumetric, there are no social tariffs and the affordability of water services is ensured through separate social policy.

Source: OECD (2007a).

Despite the reforms away from subsidised water to pricing based on supply costs and the subsequent improvements in economic efficiency, water tariffs in many cases in the OECD area remain both inefficient and inequitable. IBTs, for example, rest on the assumption that low-income households consume less water than high-income households, when in reality, poorer households can be larger, so may end up consuming more water (Sibly, 2006b; Zetland, 2011), and adjustments in their design to account for the size of large poor households cannot completely overcome the shortcomings (OECD, 2009).

Despite the increasing use of volumetric charges and the move towards covering costs, the focus of most water pricing schemes is on covering long run average costs i.e. the average cost of supplying water from existing water infrastructure (including up front and ongoing costs). The problem with such an approach is that it does not take into account the scarcity of water resources so costs do not reflect the full marginal social cost of using the resource. This leads to the inability of water prices to signal when investments in water infrastructure should be undertaken (as described in the section on water supply). Thus investments in infrastructure that determine the average cost of supply, and are made separately from the pricing decision, may be made inefficiently.

As a result of inefficient pricing in many OECD countries, imbalances between water supply and demand still arise, leading to water scarcity and the use of mandatory water restrictions. In order to resolve these problems, the optimal economic solution is to implement volumetric pricing based on the scarcity of the resource. Under such an approach, water prices rise when water is scarce and fall when water is abundant. Raising prices in times of water scarcity reduces demand (since consumers face higher prices), and increases supply (by making investments in water infrastructure profitable and raising revenues to invest in infrastructure). Thus, scarcity pricing allocates water efficiently, allowing water supply and demand to be managed to alleviate problems of water security (Grafton and Kompas, 2007; Grafton and Ward, 2010; Hughes, Hafi and Goesch, 2009; OECD, 2008; Sibly, 2006a; Zetland, 2011).

A potential concern is that scarcity pricing will lead to highly variable revenues, which may not be sufficient to cover fixed recurrent costs such as meter reading (OECD, 2009). To alleviate this problem, a fixed access fee can be included with a volumetric scarcity price to ensure stable revenues (Sibly, 2006a). However, if there are sufficient low rainfall events this may not be necessary as the revenue generated from scarcity pricing could offset potential losses when supply is at full capacity and prices are low (Grafton and Kompas, 2007). A similar concern is that this approach neglects the fact that water prices have to signal more than just volumetric water availability, such as, for example, investment and operating costs, water quality, and level of service. If volumetric pricing is introduced purely to limit consumption, it can have adverse effects by reducing revenue flows to pay for these costs. However, scarcity pricing can be implemented so that it does not base prices solely on the scarcity of water resources, rather the scarcity pricing element can be introduced in addition to existing prices. Grafton and Kompas (2007), for example, propose that water prices in Sydney should equal the short run marginal costs of supplying water when capacity is full and when storage levels decline prices should increase through a scarcity surcharge to avoid critical shortages of water.

A further concern is over the impact high water prices during times of scarcity will have on low-income households. This impact can be mitigated in a number of possible ways: 1) revenues collected during times of scarcity pricing can be used to provide water bill relief to poor households; 2) if a fixed access fee is included, it may be possible for this fee to be negative for poor households in times of water scarcity; and 3) it may be possible to introduce water use thresholds (in a similar manner to IBTs) so that poor households pay a lower volumetric charge whereas others pay the full scarcity price (Grafton and Kompas, 2007). While the use of pricing subsidies and lump-sum transfers can be an effective means of achieving equity goals, it is important that subsidies are provided independently of the level of water consumption. This is because subsidies based on the level of water consumed will cause distortions in water consumption and the allocation of water resources will no longer be efficient. Trying to achieve equity and efficiency goals with a single instrument is likely to be ineffective and better outcomes can be achieved if pricing is set so as to achieve efficiency goals and subsidies are set independently of consumption to achieve equity goals (Box 3.7).

> ### Box 3.7. **Urban water subsidies in Chile**
>
> Extensive water reforms in Chile in the 1980s led to the introduction of a new tariff for urban water prices aimed at raising prices to meet the costs of service. Prior to the reform, water tariffs covered less than 50% of costs on average and only 20% in certain regions. The reforms led to higher water prices and substantial efficiency gains, although concerns were raised over the affordability of water and sanitation services to low-income households.
>
> In order to address the equity issues, the government introduced an individual means-tested water consumption subsidy in the early 1990s. The subsidy covers 25-85% of the cost of household's basic water and sewerage consumption (up to 15 m^3 a month), with all consumption beyond this limit charged at the full price. The subsidy is targeted towards households unable to purchase the basic water needs based on a system of individual means testing. The separation of water use into two distinct goods: basic needs and optional consumption, allows the government to provide a water subsidy to low income households that is independent of water consumption beyond the basic needs.
>
> The introduction of the subsidy for basic water needs, combined with full cost pricing for further consumption, has allowed Chile to successfully raise water prices to reflect costs without compromising social and distributional goals. In 1998, nearly 450 000 subsidies were distributed, benefiting almost 13% of households by an average USD 10 per month. The cost of the subsidy scheme reached USD 42.5 million in 2000, much lower than the cost of the previous universal subsidy scheme which granted subsidies to loss making water service providers. Further, a financial deficit of 2% of assets in the water and sewerage sector was reversed to a surplus of 4% with net profits of USD 107 million, more than twice the cost of the subsidy scheme.
>
> Despite the successes, only a quarter of households in the lowest quintile of income distribution received the subsidy in 1998, suggesting that some low income households do not receive subsidies while some high income households do. In 2002, changes were made to the targeting system in order to improve the targeting of low income households.
>
> *Source:* Gomez-Lobo (2002); Gomez-Lobo and Contreras (2003).

Thus, scarcity pricing, in combination with subsidies that are independent of water consumption, is a useful market-based instrument that could be used to manage water security and to increase the efficiency, equity, and sustainability of water resources.

Water quantity

The problem of inadequate quantities of water for environmental uses is due to missing markets. While there are benefits to allocating water to the environment, these benefits are not typically represented in water markets where user's rights are traded. As a result, the benefits of environmental flows are not included in water allocation mechanisms, leading to an under-provision of water for environmental needs. Thus, the challenge of managing environmental flows is to include the benefits of leaving water in the environment into water allocation decisions that balance the trade-offs between allocating water for human use against the costs of reducing the quantity of water available for the environment.

The first-best economic solution would be to estimate the full marginal value of environmental flows in each watercourse and reach the optimal level of water abstraction, where the marginal net benefit of extracting additional water is equal to the marginal benefit of leaving it in the environment, via taxes or permits. However, because environmental benefits are typically not represented in water markets, the economic value of such benefits has to be estimated using non-market valuation techniques, such as contingent valuation, the travel cost method, and hedonic estimation (Box 3.8). Due to the intangible nature of such benefits and the complexity of implementing non-market valuation techniques, estimating the value of water in the environment is difficult and resource intensive (Van Houtven, Powers and Pattanayak, 2007; Wilson and Carpenter, 1999). Thus, the most commonly adopted approach to managing environmental flows is to define a level of water in the environment which produces acceptable environmental benefits, and reach this level through direct regulatory (command-and-control) or market-based instruments.

Box 3.8. **Non-market valuation of environmental flows in the River Murray and the Coorong, Australia**

The River Murray and the Coorong and its mouth are a unique ecosystem which provide habitat for breeding birds, fish, and vegetation. However, decreasing environmental flows during an extensive drought contributed to over-extraction and declining inflows mean that the area and its habitat have been in decline. One method of estimating the value of environmental flows is to design a survey which asks people their willingness to pay for improvements in environmental quality, using this as a measure of the value people put on the environmental services provided.

In order to estimate the value of these environmental flows in the Murray River and the Coorong, MacDonald et al. (2011) designed a survey that was sent out to over 3 000 Australian residents. The survey described the impact of low environmental flows on waterbird breeding habitat, native fish populations, and healthy vegetation in the area, and set out ways of improving environmental quality by purchasing water user's rights from willing sellers, investments in irrigation efficiency, and habitat rehabilitation, together with the costs of these policies. The survey then asked respondents to choose between various policy options that had different environmental impacts and different costs.

Through a statistical analysis of the results from the survey, MacDonald et al. (2011) found that Australian residents were willing to pay substantial amounts to improve the quality of the Murray River and Coorong indicating that the value of environmental flows in the area is significant. Specifically, total willingness to pay (in present value terms) to increase the frequency of waterbird breeding from every 10 years to 4 years, to increase native fish populations from 30 to 50% of original levels, to increase the area of healthy native vegetation from 50 to 70%, and to improve waterbird breeding habitat quality in the Coorong was AUD 13 (USD 14) billion. The authors stress that, due to the uniqueness of the Coorong, this value cannot be used to estimate the value of other watercourses in Australia, and further surveys are required.

Source: MacDonald et al. (2011).

The two most developed systems for managing environmental flows are those in South Africa and Australia, where direct regulatory and market-based instruments have been adopted respectively.[8] In South Africa, the 1998 National Water Act overhauled the existing riparian rights based system of water law and introduced a system centred on providing equitable water user's rights. The new law established the provision of water for basic human and environmental needs as the only recognised legal rights of water users, with all other uses requiring entitlements that are only granted if the use is beneficial to the public. The amount of water set aside for the environment, or the ecological reserve, is determined by scientific assessment of the requirements to achieve ecological sustainability. The ecological reserve applies to both surface and groundwater and the quantity of water required by the reserve must be met before any other water use permits are allocated (Hirji and Davis, 2009).

While the scope and ambition of water reform in South Africa is considerable, in practice the success has so far been limited. This is because water users' rights for most water resources were already fully allocated under the previous riparian system, and as a result, the introduction of the ecological reserve requires a reduction in water allocations for users who have held entitlements for long periods of time. Further, since the Act does not advocate the use of market-based instruments, instead favouring a regulatory approach to the reallocation of water, entitlement holders have to give up their user's rights without compensation. This has created considerable opposition to water reallocation as users have no incentive to give up their user's rights, and so has delayed the successful introduction of ecological reserves (Andreen, 2011; Hirji and Davis, 2009). A wider problem with the South African approach to securing environmental flows is the use of command and control approaches. As command and control methods do not address the incentives of water users to use water sustainably, nor compensate water users for reductions in water use, they can generate significant political opposition, particularly from politicised bodies such as farmers groups. Such approaches are also more vulnerable to distortion, extortion, and corruption, making the implementation of command and control approaches politically difficult, as demonstrated in the case of South Africa.

By contrast, the Australian approach to the management of environmental flows uses market-based instruments in order to address the incentives of those who have to give up water user's rights. In the Murray-Darling Basin, low environmental flows had significant negative impacts on the environmental health of the region with 20 of the 23 river valleys in the Basin classified as in poor or very poor health during an extensive drought (Grafton, 2010; Pittock and Finlayson, 2011). In order to increase the level of environmental flows to sustainable levels, a Basin Plan determined by an independent authority established by the Australian government set sustainable diversion limits (SDLs) that place a maximum limit on the quantity of water which can be taken from the Basin. Recent scientific modelling suggests that ensuring environmental sustainability requires at least 60% of natural flows remain for environmental needs; which would involve reducing the current limit on diversions by some 30-40% (MDBA, 2011; Wentworth Group of Concerned Scientists, 2010).

In order to achieve these reductions, the Australian government is currently using two market-based instruments: 1) infrastructure subsidies, including upgrades to public infrastructure and improvements in on-farm water use efficiency, with half of the water savings resulting from efficiency improvements allocated towards environmental flows; and 2) market-based recovery, through buy-backs of water user's rights in the water market and through a tender process. To date, around 1 200 GL has been acquired for the environment (Wentworth Group of Concerned Scientists, 2010).

Research suggests that, compared to the buy-back scheme, the infrastructure subsidy scheme is both cost-inefficient and environmentally ineffective. Qureshi et al. (2010) estimated that the infrastructure subsidy scheme implemented on the Murrumbidgee River in the Murray-Darling Basin could secure a maximum of 143 GL of water for the environment at a cost of up to AUD 6 000 (USD 5 160) per GL, whereas the buy-back scheme could secure an additional 733 GL of environmental flows at AUD 3 000 (USD 2 580) per GL. In addition, the subsidy scheme has been criticised on grounds that the lag time before environmental flows are secured creates risks to already vulnerable ecosystems, and investing in water infrastructure may lock communities into an irrigation-dependent future, when in some cases it may not be viable due to climate change and further water buy-backs (Wentworth Group of Concerned Scientists, 2010).

The key reason why the buy-back of water user's rights is more cost-effective is that it allows for flexibility in the way water reductions are achieved. Under the buy-back scheme farmers can reduce their water use in the least-cost manner, such as through deficit irrigation, changing land use, or through improvements in irrigation efficiency. In the subsidy scheme, however, reductions in water use are only achieved through improvements in irrigation efficiency, whether or not this is the least-cost method (Grafton, 2010). The reduction of non-consumptive water uses, such as reducing irrigation leakage via infrastructure subsidies, can also lower the amount of return flows to the environment and, as a result, increases in irrigation efficiency may actually reduce the net level of water which is available for the environment (Qureshi et al., 2010). Further, the economic impact of water reductions on agriculture associated with the buy-back of water user's rights are fully compensated for and trading in water markets allows water to be transferred from low to higher value uses so that when water diversions are reduced, the least profitable uses are reduced first (Grafton and Jiang, 2010).

A potential limitation with the buy-back scheme is that it depends on seller participation in the market. Wheeler et al. (2010) found there was considerable reluctance among irrigators in the Murray-Darling Basin towards selling water to the government and estimated that the total volume of water that irrigators were willing to sell was only around half of that needed to be acquired at existing prices. A potential solution to this problem is to slow the acquisition of entitlements over a longer period of time and incorporate the purchase of temporary allocations which irrigators are more willing to sell at existing market prices (Productivity Commission, 2010; Wheeler et al., 2011). Such an approach could potentially increase the willingness of irrigators to participate; increase the amount of water available for environmental flows; and allow a longer period of adaption for irrigators and communities, thereby lowering the socio-economic costs of securing environmental flows (Wheeler et al., 2011).

The socio-economic costs of acquiring environmental flows through water markets raise further concerns over the equity of the buy-back scheme and the impact buy-backs will have on regional communities that are dependent on irrigated agriculture. However, studies estimate that, due to the flexibility of water markets, substantial reductions in water extractions will have only a moderate impact on net profits from irrigated agriculture, and employment in the Basin may actually increase due to the stimulus impact of the buy-backs combined with subsidies (ABARE, 2010; Grafton and Jiang, 2011). Although the overall socio-economic costs of securing environmental flows are likely to be modest, the flexibility of the buy-back approach may concentrate reductions in water extractions in a few specific areas and a few less profitable crops (Grafton and Jiang, 2011). As a result, small towns which are heavily dependent on irrigated agriculture in these areas may be significantly affected (ABARE, 2010). Easing the transition to reduced water extractions in such areas is therefore an important issue. However, using the buy-back of water user's rights so as to achieve distributional goals is likely to reduce its efficiency and effectiveness and so equity goals are best addressed directly with other instruments (Productivity Commission, 2010).

In sum, the management of environmental flows in South Africa and Australia demonstrate the advantages that market-based instruments have over direct regulatory instruments in terms of making reallocations by addressing the incentives of entitlement holders and reducing water allocations in a cost-efficient manner. The experience of both countries also suggests that flows should be allocated at a river level rather than nationally because the environmental needs of different river systems differs. Indeed, it may be possible to have some rivers with very few environmental flows if they are offset by higher flows in other rivers. In any case, environmental flows should be based on desired environmental outcomes for each river system such that the use of national minimum flows, as opposed to targets for specific rivers, is inappropriate.

Water quality

For the most part, water quality in the OECD has been managed through direct regulatory instruments such as setting ambient water quality standards, technology requirements, controls on polluter's emissions into sewer systems and water courses, and bans on discharges into water sources used for drinking or irrigation.

Due to the increasing costs required to meet tougher water standards and past implementation of the cheapest and easiest ways to achieve pollution reductions, there is a growing shift towards the use of market-based instruments such as emission taxes and emission trading, which can theoretically achieve the desired water quality standards at much lower cost (Kraemer, Kampa and Interwies, 2003).

The use of market-based instruments to resolve water quality problems is based on the concept of externalities; pollution leads to declining water quality which imposes costs, or negative externalities, on society and the environment. The problem of water pollution, therefore, arises because the full cost of declining water quality is not borne by the polluter, driving a wedge between the private costs of discharging pollutants and the social costs they impose. Since polluters do not bear the full marginal cost of their actions, this leads to excessive water pollution.

An economic solution to this problem is to impose a Pigouvian tax on each unit of pollution which is equal to the marginal social cost of pollution at the optimal level (Figure 3.1). This requires that planners know the optimal level of pollution and the marginal social costs of each additional unit of pollution. While this can be estimated through a number of techniques, such as contingent valuation, the travel cost method, and hedonic estimation, consistent estimates of the full costs of pollution are often difficult to find (Van Houtven, Powers and Pattanayak, 2007; Wilson and Carpenter, 1999). Thus, instead of aiming to achieve the optimal level of pollution, planners often define an acceptable level and use taxes to reach it at the lowest possible cost (Tietenberg, 1990).

An alternative economic solution is to set a maximum limit on the emissions of a particular pollutant in a watercourse, and allocate the total amount of emissions among polluters through tradable emission permits (Figure 3.1). In either case, a cost minimising polluter will seek to minimise the sum of acquiring emission permits or paying emission taxes, and the cost of pollution abatement. The cost minimising point for each polluter will occur where the marginal cost of abatement is equal to the price of the emission permit or the emission tax. Since each cost minimising polluter abates until the marginal cost of

Figure 3.1. **Reaching the optimal level of pollution via emissions taxes and emissions permits**

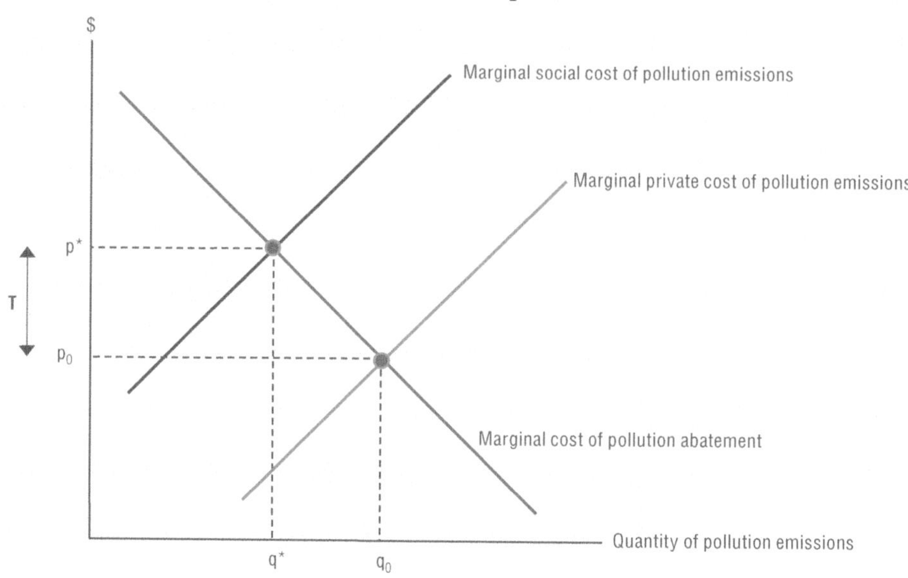

Note: The optimal level of pollution is where the marginal social cost of pollution emissions (increasing with emissions) is equal to the marginal cost of pollution abatement (decreasing with emissions). To reach this point a Pigouvian tax (T) can be charged on each unit of pollution, raising the marginal private cost of pollution to that of the marginal social cost. Alternatively, a fixed number of tradable pollution permits can be issued at the optimal level and can be traded among the sources of the emissions of pollution (q^*) to generate a market price for emission permits equal to p^*.

Source: Adapted from Tietenberg (1990).

abatement is equal to the permit or tax price, marginal abatement costs are equalised across all polluters and the outcome is cost-efficient (Tietenberg, 1990).

Market-based instruments are theoretically more cost-effective than direct regulatory instruments which impose the same controls on all polluters and do not take into account the heterogeneity of abatement costs (Tietenberg, 1990). Market-based instruments also provide a dynamic incentive for additional pollution abatement, as polluters can reduce their costs by the amount of the emission tax or permit price for each additional unit of pollution abatement. This incentive effect can lead to significant investment in pollution abatement and technological innovation thereby lowering the overall cost to society of meeting environmental targets, an outcome that may not be realised under direct regulatory instruments where there is no incentive for abatement beyond that which is required (OECD, 2010b). A further advantage of market-based instruments is the potential of a double dividend: first, by leading to environmental improvements; and second, by raising revenues for the government which can be used to reduce distortionary taxes thereby creating further efficiency gains, such as the United Kingdom government's use of revenue from waste taxation to reduce employer's social security contributions (Cowan, 1998). Whether or not this double dividend can be achieved in practice is a matter of debate, however, even without a double dividend, revenues raised from market-based instruments can be used to offset the direct impacts of the tax (OECD, 2006; 2010a).

Despite these advantages, the adoption of market-based instruments to manage water quality has been relatively slow in many parts of the OECD. One reason for this is the complexity of water pollution problems. Unlike air pollution, where emissions mix into the atmosphere regardless where they are released, water pollution is location-specific and emissions damage the specific watercourse they are released into. Thus, the marginal costs of water pollution vary dramatically with the location of emissions making it difficult to design cost-effective policies (Olmstead, 2010). As a result, a uniform emissions tax across different watercourses may actually lower welfare if the difference in marginal damages is significant (Cowan, 1998).

In theory, emissions taxes should vary according to each watercourse and discharges in environments with a high diluting capacity (such as offshore) should be charged less than in areas with low diluting capacity as the marginal damage of additional discharges is much lower. Thus, the level of tax for units of pollution should be defined according to the quality of the recipient body, and ideally emissions taxes should be imposed based on the ambient standards of each body they are discharged into. However, due to the difficulty of measuring ambient water quality for each water course, the cost of finding the optimal level to charge in each case and the regulatory complexity of implementing differentiated taxes, standardised unit pollution taxes are often implemented. Likewise, the difference in marginal damages of emissions across watercourses means that emissions trading schemes are usually limited to individual watercourses. Further, trading within a watercourse may lead to a concentration of permits in one area and the cumulative impact of pollution may lead to greater ecological damage. These problems create significant administrative, monitoring, and enforcement costs for planners, limiting trade and reducing efficiency (Boyd, Shabman and Stephenson, 2007).

Market-based instruments are, therefore, best implemented in a location-specific context. Emissions permits, for example, may be preferable when the marginal damages of pollution in a watercourse are relatively steep, since taxes have an unknown impact on emissions, whereas permits are set a fixed level. Emission taxes on the other hand may be preferable when there are few polluters in a watercourse and markets for emission permits are likely to be uncompetitive and subject to high transaction costs.

In practice, emissions taxes have been used in a number of countries for two purposes: 1) reducing water pollution; and 2) raising revenues. In the Netherlands, for instance, emissions taxes were set at a very high level in 1970 which led to a reduction in total organic emissions by 50% and industrial organic emissions by 75% by 1990 (Stavins, 2003). Likewise, high emissions taxes have been implemented in Germany, the Czech Republic, and Slovenia, in order to encourage behavioural change and reduce water pollution (Peszko and Lenain, 2001). However, in the majority of countries where emissions taxes have been implemented, they have been set at too low a level to induce behavioural change, and so have primarily been used to raise revenues (Glachant, 2002; Peszko and Lenain, 2001; Stavins, 2003). Across the OECD area, there is increasing use of separate emissions taxes in industrial and urban water bills (OECD, 2010b). In most cases, emissions taxes were initially implemented to cover costs, however, they are increasingly being used to provide incentives for users to continuously reduce discharges and in some cases they can be significant. In France, for example, emission taxes now make up around 12.5% of household water bills (Bommelaer et al., 2011).

The use of emissions trading schemes is also increasing although they are less common than emissions taxes. To date, most trading schemes have been implemented in the United States and Australia and their success has been mixed (Boyd, Shabman and Stephenson, 2007; Kraemer, Kampa and Interwies, 2003; Selman et al., 2009). Successful schemes include the Long Island Sound Nitrogen Credit Exchange Programme in the

United States, where 12 million credits have been bought or sold at a value of USD 30 million (Selman et al., 2009); and the Hunter River Salinity Trading scheme in Australia, where the salinity target has not been exceeded due to polluter's discharges since the scheme has been in operation, and water treatment and storage costs have been significantly reduced (Kraemer, Kampa and Interwies, 2003). The success of these schemes is due to the flexibility allowed to firms due to minimal regulation on trades, the large number of eligible participants in trading markets, effective monitoring and enforcement procedures, and strong legislation underpinning the schemes.

In a number of schemes where there were few polluters and extensive administrative requirements were placed on trade, such as the organic pollutants trading scheme in the Fox River (United States), markets were less successful due to higher transaction costs, lower gains from trade, and less incentive to participate (Kraemer, Kampa and Interwies, 2003). The difficulty of implementing emissions trading schemes has led to the use of offset schemes in place of pure trading schemes, such as in the Cherry Creek and Rahr-Malting rivers in the United States (Boyd, Shabman and Stephenson, 2007). Offset programmes allow polluters to offset their discharges by purchasing emissions abatements from other polluters or by investing in pollution abating projects such as wetland restoration. However, unlike pure emissions trading schemes, the flexibility rests with regulators rather than polluters since each trade requires *ex ante* approval from the regulator, polluters cannot reduce their control levels by offsetting increases in other polluter's control levels, and there is no overall cap on emissions. While offset schemes are easier to implement due to less legal, administrative, and technical complexity, they are less flexible and so are less efficient than pure emissions trading (Boyd, Shabman and Stephenson, 2007). Thus, pure emissions trading schemes promise much greater benefits provided they can be implemented successfully.

The combination of emissions standards, taxes, and trading in the OECD has been largely successful at reducing point source pollution, particularly from urban and industrial sectors. However, little progress has been made in tackling non-point sources of pollution, primarily from agricultural sources, as they are much more difficult to manage (OECD, 2008).

The key challenges of dealing with non-point sources of pollution are: first, it is much more difficult to identify and monitor the actual sources of pollution; and secondly, that ambient levels of non-point source pollution are influenced by the weather and other environmental factors so have a strong stochastic element (Olmstead, 2010). Being unable to target specific sources of emissions raises the question of what pollution control measures should one target, and a variety of solutions have been proposed such as the difference in nutrient inputs to outputs across farms (Box 3.9); particular inputs or practices which are associated with pollution; and ambient levels of pollution (Shortle and Horan, 2001).

An OECD report outlines a number of market-based instruments that have been proposed to tackle non-point sources of pollution due to the failure of direct regulatory instruments such as design and performance standards (OECD, 2007b). These include taxing inputs which are associated with water pollution, such as fertilisers and pesticides; although this may encourage animal production over vegetable production, leading to an increase in manure levels (OECD, 2007b). Alternatively, taxes can be placed on ambient pollution levels so that farmers are charged if ambient levels are above a certain threshold and receive a subsidy if below. While this approach reduces the monitoring costs involved, there are problems with implementation such that farms that reduce their own levels of pollution may

> ### Box 3.9. **MINAS accounting system in the Netherlands**
>
> The nitrogen (N) and phosphorus (P) accounting system (MINAS) was introduced in the Netherlands in 1998 in order to tackle non-point sources of pollution from farms and represented a shift from ineffective direct regulatory instruments towards market-based instruments.
>
> The MINAS system used the difference between N and P inputs and outputs at a farm level as a proxy for non-point source emissions, requiring the registration of all N and P inputs (fertilisers, manure, and animal feed) and outputs (export in harvested products) for each farm. An acceptable level of surplus N and P was determined (units of N and P relative to the surface area of the farm) and if the farm's difference exceeded this surplus, they had to pay a levy proportional to the excess above the surplus.
>
> MINAS gave farmers the option of continuing to pollute and paying the levy, or reducing their pollution to avoid it. In order to encourage farmers to choose pollution reduction, the acceptable level of surplus decreased from 1998 to 2003 and the size of the levy increased, so that by 2002 the levy was about 5-10 times the price of fertiliser N and 50 times the price of fertiliser P.
>
> The policy was successful at significantly reducing N and P pollution from dairy farms. However, it was not successful for pig or poultry farms, which had higher inputs of animal feed and smaller farm surface areas. It was also unsuccessful for arable and horticultural farms because P pollution was excluded and the acceptable level of surplus was set too high. Further, the administration and enforcement costs increased rapidly due to fraud, exploitation of loopholes, and increasing changes made to the system, meaning that the costs were much higher than other potential policies. Thus, in 2005 it was replaced by an alternative system.
>
> Although the MINAS system was a promising instrument, it failed to fully account for the considerable heterogeneity across farms in the Netherlands and the implementation period was not long enough to allow it to be fully fine tuned.
>
> *Source:* OECD (2006; 2007b).

still be taxed if others increase their levels of pollution by a greater amount, or ambient pollution levels rise due to unforeseen weather or other environmental factors (Dowd, Press and Huertos, 2008). An alternative approach is to issue payments to farmers to adopt specific inputs or practices which reduce pollution, although monitoring and enforcement costs are again significant and the adoption of best practice management may not translate into improvements in ambient water quality (Dowd, Press and Huertos, 2008).

A further market-based instrument which is receiving increasing attention is the use of point-non-point source emissions trading schemes; where point source polluters can reduce their pollution abatement requirements by purchasing an equivalent amount of pollution abatement from non-point sources (Olmstead, 2010). The widening gap in marginal abatement costs between point and non-point sources means that such trade could reduce the costs of improving water quality significantly. The US Environmental Protection Agency, for example, estimates that trading between point and non-point sources could lower the costs of the Total Maximum Daily Load Programme by up to USD 235 million each year (US EPA, 2001).

Complications arise because point source emissions cannot be traded directly with non-point emissions as these are unobservable and stochastic. Point-non-point trading schemes, therefore, involve trading of point source emissions for reductions in the use of inputs that are correlated with non-point source pollution, or reductions in estimated loadings from non-point sources. In either case, reductions in inputs or estimated loadings are imperfect substitutes for point source emissions so they should not be traded on a one to one basis (Shortle and Horan, 2001). The Lake Dillon Trading Programme in Colorado, for instance, allows non-point source credits to be traded for point source credits at a 2:1 ratio so that point source polluters have to reduce non-point sources of phosphorous pollution by two tonnes before they can increase their discharge by one tonne (Kraemer, Kampa and Interwies, 2003).

With its Lake Taupo nitrogen trading scheme, New Zealand introduced the first non-point source to non-point source (NPS) cap and trade scheme worldwide (Shortle, 2012). Despite the importance of NPS pollution worldwide, to date, water quality trading markets have predominantly been set up to facilitate nutrient discharge reductions by point sources, such as sewerage plants and mines. Where agricultural NPSs are involved, they are generally not subject to a cap on emissions, and instead can choose to participate and decrease nutrient discharges in return for emission reduction credits that point sources purchase to offset their own discharges (Selman et al., 2009). The Taupo scheme is innovative as controlling diffuse NPS nutrient discharges is its central aim. Young et al. (2010) and Duhon et al. (2012) have discussed the process of creating the system, and evaluated its early operation (Box 3.10).

Box 3.10. **Two case studies of water quality trading: Lakes Taupo and Rotorua (New Zealand)**

Lake Taupo is New Zealand's largest lake with a catchment of nearly 3 500 km^2 of pastoral farms, plantation forestry, native forest and a small urban area. The lake has been described as "iconic". It is a major destination for domestic and international tourism. Although Lake Taupo currently exhibits exceptional water quality, scientific investigation has revealed a gradual but steady decline in key indicators of water quality over the past three decades (Vant, 2008). Intensified pastoral and urban land use over the past 35-50 years has resulted in increased nutrient levels in the lake, leading to decreasing water quality and clarity (Young, 2007). Water quality is expected to decline further even if current discharge levels are capped because of considerable time lags in the Lake Taupo catchment between nutrient application to land and its eventual arrival in the lake via groundwater. This time lag ("latent risk") is thought to be greater than 100 years in some parts of the catchment (Vant, 2008; Hadfield, 2008).

Nitrogen losses from agricultural land uses have been identified as the primary cause of increased nutrient loads into the lake. Total nitrogen discharges into the lake are around 1 360 tonnes per year, of which only 556 tonnes per year come from manageable or human-induced sources. Pastoral (dairy and sheep beef) activities account for 92% of all manageable sources of nitrogen loss.

Lake Taupo is located in the Waikato region. Following growing community concern about water quality, Waikato Regional Council set a goal to restore water quality to 2001 levels by the year 2080. This long time horizon reflects the latent character of the water pollution risk. Under New Zealand Resource Management Act (1991), the Regional Council is responsible for water quality (Kerr et al., 1998). The policy designed to achieve this goal consists of three key components: a cap, a public fund for buy-backs, and trading. The catchment level cap on nitrogen losses serves to limit nitrogen losses at historical levels and prevent further increases ("acceptable level of risk"). A computer model OVERSEER is used to model leaching from each of the roughly 250 farm participants based on auditable data. Each farm is benchmarked to

> ### Box 3.10. **Two case studies of water quality trading:**
> ### **Lakes Taupo and Rotorua (New Zealand)** (*cont.*)
>
> initially grandparent allowances and then must comply with a management plan to ensure compliance. The Lake Taupo Protection Trust, a public fund with contributions from local, regional and national communities, is charged with permanently reducing the catchment cap by 20% through the purchase and conversion of land or purchase and permanent retirement of farmers' nitrogen allowances. The nitrogen trading system allows farmers to trade allowances with other farmers or with the Trust. To make a trade, both the buyer and seller must submit an updated nitrogen management plan for Council approval.
>
> The policy became fully operative in July 2011 after resolution of some legal challenges but trades had been being negotiated since 2007 when the Lake Taupo Protection Trust was given the ability to make nitrogen discharge allowance (NDA) purchase decisions (Young et al., 2010). The first Trust and private trades were completed in January 2009.
>
> In contrast, Lake Rotorua does not yet have a nutrient trading system. It has had a weakly monitored freeze on leaching from each farm since 2005 and active negotiations on more stringent rules are occurring at catchment scale among landowners and other local stakeholders and with the regional government. Rotorua is interesting because it has more intensive land use and a more severe water quality problem than Taupo and because it is one of 16 lakes in the Rotorua area (Bay of Plenty region) but the only one where nutrient trading is likely to be part of the solution. It offers an opportunity to learn from the Lake Taupo experience and refine the nutrient trading model even further. It has been the location of considerable policy and integrated modelling research.[1] One issue of specific interest is the role of groundwater lags. It is estimated that 53% of the nitrogen reaches the lake via groundwater with lags up to 120 years (Rutherford et al., 2011).
>
> The key differences between the Taupo trading system and that being proposed for Rotorua (Kerr et al., 2012), are that the Rotorua system attempts to avoid a need for approved farm management plans or Regional Council approval for trades by using a self reporting system; that more certain and swift non-compliance penalties are being explored (Rive, 2012); and that initial allocation of allowances is likely to be less generous to farmers. These all reflect both learning from the Taupo experience and the need for more stringent reductions. Allocation may also be done on a different basis reflecting concerns about the fairness of grandparenting, particularly for Maori landowners who regained control of their land only under recent Treaty of Waitangi settlements, and those who have undertaken voluntary mitigation.
>
> 1. For a range of papers on nutrient trading for Lake Rotorua, see *www.motu.org.nz/research/detail/nutrient_trading*.

Despite the increasing use of market-based instruments to tackle non-point source pollutions, there has so far been very little empirical assessment of the cost-effectiveness and environmental success of such schemes. While market-based instruments are potentially a very promising solution to tackling non-point sources of pollution, they are still in a stage of development and a mixed approach involving a combination of instruments may be the best step at the present time (Dowd, Press and Huertos, 2008). Further, while the use of market-based instruments could lower the costs of achieving pollution reductions, the complexity of non-point pollution problems mean that it may not be possible to efficiently address all aspects with the same type of instrument. Thus, there is a strong argument for using a combination of instruments to tackle non-point sources of pollution including command and control measures, market-based instruments, as well as wider agricultural sector reform such as reducing subsidies which encourage intensive agriculture, or introducing cross compliance measures so that farmers have to comply with environmental regulations before they receive subsidies (OECD, 2007b).

Integrated water resource management

In order to deal with the complexity of water security management, this analysis separated water security into four separate issues and focused on how market-based instruments could be applied to provide efficient, cost-effective solutions to each particular issue (see Table 3.1 above). In reality, each of these issues are interconnected; rising water demand, for example, may reduce the quantity of water in the environment that can increase the concentration of pollutants that may lead to a decline in water quality. Thus, the use of market-based instruments to manage one issue affects the management of others, and co-ordinated use of market-based instruments can increase the cost-effectiveness and efficiency of water security management as a whole. Introducing scarcity pricing, for instance, will affect both water supply and water demand by reducing demand for water and signalling the optimal time to invest in water supply augmentation.

Due to the interrelated nature of water security issues, the use of market-based instruments may create trade-offs. Thus, while market-based instruments can create efficient, cost-effective solutions at a local or basin level, they may also have wider social or environmental impacts. Purchasing water user's rights to secure environmental flows, for example, may be a cost-effective method of increasing environmental flows, but reduced water extractions may also negatively impact small towns and communities dependent on irrigated agriculture. It is also possible that although the trading of water pollution permits is an efficient method of reducing overall pollution, in the absence of regional trading constraints it may concentrate pollution permits in particular areas leading to pollution hotspots.

Thus, for water security to be managed effectively, the use of market-based instruments cannot be considered in isolation, but rather must be considered in terms of their wider impact on society and the environment. Effective water security management, therefore, requires planners to take into account the "triple bottom line" and evaluate policies in terms of their economic, environmental, and social impacts. In order to deal with this complexity and coordinate policy effectively, market-based instruments must therefore be used as part of a wider integrated water resource management (IWRM) framework.

IWRM is a framework designed to improve the management of water resources based on four key principles which were adopted at the 1992 Dublin Conference on Water and the Rio de Janeiro Summit on Sustainable Development. These principles hold that: 1) fresh water is a finite and vulnerable resource essential to sustain life, development, and the environment; 2) water development and management should be based on a participatory approach, involving users, planners, and policy makers at all levels; 3) women play a central part in the provision, management, and safeguarding of water; and 4) water has an economic value in all its competing uses and should be recognised as an economic good (ICWE, 1992).

IWRM is not a prescriptive description of how water security should be managed, but rather it is a broad framework in which decision makers can collaboratively decide the goals of water security management and co-ordinate the use of different instruments to achieve them (Lenton and Muller, 2009). Given that each country differs in terms of their history, socio-economic conditions, cultural and political context, and environmental characteristics, there is no single blueprint for IWRM and it should be adapted to resolve the problems faced in each local context (Pahl-Wostl, Jeffrey and Sendzimir, 2011).

The goals of IWRM vary across countries and different weights are placed on the importance of economic, environmental, and social impacts: Chile, for instance, emphasises the importance of economic efficiency, whereas South Africa and the Netherlands tend to place more emphasis on social and environmental goals respectively. It should not, however, be thought that there are always trade-offs between these goals, and a more integrated approach to water security management can help in achieving win-win outcomes which promote more than one goal. Implementing scarcity pricing, for example, promotes economic efficiency, creates environmental benefits due to decreases in water demand, and can generate social benefits if combined with subsidy or rebate schemes for low-income households.

Despite the differences in implementation across countries, IWRM can be broadly characterised by a number of key trends. *Firstly* (and the focus of this review), there has been a move away from direct regulatory instruments which focus on supply-side water management, such as large-scale water infrastructure, towards incorporating demand side management, though the use of market-based instruments. This shift in focus has created a more flexible approach to water security and has encouraged the development of a variety of innovative market-based instruments to resolve local water security problems. *Secondly*, IWRM has led to an increased awareness of the importance of sustainable development and the incorporation of social and environmental considerations into water management. In many cases this awareness has led to these issues being effectively addressed through the IWRM framework. *Finally*, IWRM has led to a move away from top-down, centralised approaches to water security towards more flexible, decentralised approaches which involve a variety of diversified governance structures at a local, basin, national, and transnational level (Lenton and Muller, 2009). Basin-level management, in particular, is critically important to good water outcomes given the lack of mobility of water and the need to design market-based instruments that adapt to specific basin conditions (Box 3.11).

Funding for basin governance can be designed in various ways, such as through general revenues, higher water prices, water abstraction charges, etc. Ideally, the funding model should be designed in a way that distorts economic outcomes the least. Thus, ideal funding mechanisms will vary by basin and country. Finally, there is increasing stakeholder collaboration and the involvement of local communities in water security decision-making. The benefits of wider collaboration include: incorporating specialised knowledge; encouraging more innovative solutions to problems due to greater diversity of viewpoints; encouraging co-operation and reducing the risk of conflicts over water resources; and developing solutions which are more open, inclusive and democratic, thereby generating wider support and leading to more sustainable outcomes (Loux, 2011).

There are, however, challenges which IWRM faces: the lack of a clear, prescriptive definition means that it is often difficult to implement; collaboration is often time-consuming and resource intensive; the level of co-ordination required for large projects may make IWRM too complex to undertake, particularly for developing countries which lack the necessary institutions; and the flexibility of implementation means that it is difficult to evaluate the performance of IWRM itself compared to the particular choice of instruments (Biswas, 2008; Loux, 2011; Pahl-Wostl, Jeffrey and Sendzimir, 2011). Nevertheless, there is growing evidence, and particularly within OECD countries, that implementing IWRM can offer substantial, long-term benefits to water security and water management (Lenton and Muller, 2009; Pahl-Wostl, Jeffrey and Sendzimir, 2011).

> ### Box 3.11. **IWRM in the Lerma-Chapala River Basin, Mexico**
>
> The Lerma-Chapala River Basin is one of world's most water-stressed basins. Rapid population growth combined with industrial and agriculture development have led to serious imbalances between water withdrawals and availability. Further, the increasing competition over water resources in the basin, combined with poor governance, has resulted in over-exploitation of surface and ground water resources, increasingly frequent conflicts over water allocations, and considerable levels of water pollution and soil degradation. As a result, during the period from 1981 to 2001, Lake Chapala lost 90% of its natural volume and the remaining water was left heavily contaminated.
>
> Recently, however, due to a move towards IWRM and subsequent improvements in water governance, the situation has begun to improve and: the natural capacity of the lake has been restored; water quality is improving with around 60% of discharges eliminated; irrigation efficiency has risen; and finance has been secured to invest in water sanitation and treatment programs.
>
> The improvement in water governance is due to reforms beginning in the 1970s which started a move away from centralised governance in Mexico towards IWRM. By the early 1980s, six regional water resources offices were set up, including the newly created Lerma-Chapala River Basin Regional Management agency which was given the responsibility of gathering information and designing a Basin Plan. Further reforms in 1992 and 2004 strengthened the decentralisation process and set up Basin Councils with formal powers to implement the proposed water reallocation policies.
>
> The Lerma-Chapala Basin Council carried out a hydrological study of the Basin and developed a model to evaluate the impact of various water reallocation policies according to economic, social, technical, political, and environmental criteria. This model was then used as a basis for water reform in the Basin. The Council also encouraged extensive collaboration with stakeholders in the Basin and took steps to communicate their work as transparently as possible which reduced the level of conflict over reallocations.
>
> While the move towards IWRM in the Lerma-Chapala Basin has been a long and difficult process, after 30 years, the benefits are starting to be realised.
>
> *Source:* Hidalgo and Pena (2009).

Conclusions

In terms of water supply, market-based instruments can be used to efficiently augment the natural water supply by making efficient investments in water infrastructure and, when appropriate, through the purchase of water via inter-basin transfers. In particular, the use of scarcity pricing of water resources can be used to signal the optimal time to invest in large-scale water infrastructure projects, thereby, avoiding the considerable welfare losses associated with water prices being raised to cover the costs of poorly timed investments. Similarly, vertical separation (unbundling) of bulk water supply from retail distribution and also horizontal separation of the retail and waste-water sectors offer the potential of efficiency gains.

With respect to water demand, market-based instruments can be used to efficiently and effectively manage demand through water markets and water pricing. Water markets create incentives to reduce water demand and allow the transfer of water to its highest valued uses, ensuring that the resource is allocated efficiently. Evidence suggests that well-designed water markets, such as those in the Murray-Darling Basin in Australia, can have significant economic benefits as well as reducing water demand. Likewise, dynamically efficient water pricing that incorporates the scarcity value of water resources can be used to manage demand efficiently, effectively, and equitably if combined with subsidies or rebates to low-income households. Thus, scarcity pricing can be used to manage both water demand and water supply.

For water quantity, market-based instruments can be used to secure water for environmental flows in the most efficient manner. Where water markets are operating, buy-backs of water user's rights through markets can be used to secure environmental flows efficiently by purchasing the lowest value uses of water first. Further, the compensation received by sellers of water user's rights, combined with the flexibility created by water markets, reduces the economic impact of reductions in water use. Thus, water markets can be used to manage both water demand and water quantity. For areas where water is not already over-allocated, water allocation limits should be set after assessing the risk to environmental, cultural and social water-dependent values, as well as the development risks of not abstracting water for consumptive use.

Market-based instruments can be used to help manage water quality, in conjunction with other policy instruments, to reduce the costs of improving water quality. For single identifiable ("point") sources of pollution, the combination of emissions standards, taxes and trading has created significant improvements in water quality across the OECD, and the increasing emphasis on market-based instruments has reduced costs and encouraged technological innovation. For non-point sources there has been less success, however, the use of point-non-point emission trading schemes and a variety of other market-based instruments in conjunction with emissions standards may be effective at reducing non-point source emissions. With its Lake Taupo nitrogen trading scheme, New Zealand recently introduced the first non-point-source to non-point-source cap and trade scheme worldwide.

As water risks are interlinked and the use of market-based instruments can have wider environmental and social impacts, a focus on economic efficiency by itself is not sufficient to tackle water security problems. Environmental and social goals need also to be considered. A widely accepted framework to implement this integrated approach is through integrated water resource management (IWRM), which encourages a more flexible, adaptive approach to water security management, involving greater collaboration with stakeholders and increasing the chance of sustainable outcomes to water security problems in the long term.

Notes

1. An *adequate* supply for essential personal and domestic uses, which include drinking, sanitation, washing of clothes, food preparation, and personal and household hygiene, requires a minimum of 50 to 100 Litres per person per day (Howard and Bartram, 2003).

2. UN Resolution 64/292: The Human Right to Water and Sanitation (30 September 2010). Numerous subsequent UN Resolutions have been issued.

3. International trading of bottled water is significant and growing (Gleick et al., 2002), however, the small volumes and prohibitively high prices of bottled water mean that it is not a substitute for water in industrial, agricultural, or urban use. Therefore, in this analysis, bottled water is treated as a separate product from raw water, and its trade is not discussed.

4. Since being signed in 1997 the Convention on the Law of the Non-navigational Uses of International Watercourses is not yet in force.

5. Except for Giffen and Veblen, goods such that an increase in price can lead to higher demand. Basic goods, such as rice, may be a Giffen good for some households because the income effect outweighs the substitution effect (Battalio, Kagel and Kogut, 1991). For example, as the price of rice rises low-income households have less money to spend on more expensive alternatives, such as meat, and so may consume more rice to make up for the decrease in amount of meat consumed. Expensive or luxury goods, such as luxury cars, may be Veblen goods for some consumers because the good's use represents conspicuous consumption. For such goods the higher is its price, the more status the good confers and, hence, the greater is the demand (Veblen, 1899).

6. Therefore, a 10% increase in the average water price across households would lower urban water use by about 5.6%.

7. The use of flat fees, however, is still reported in Canada, Mexico, New Zealand, Norway and the United Kingdom.

8. While the United States has a long history of managing environmental flows, the US system is highly fragmented with no comprehensive approach to maintaining flows and the significant gaps in regulatory authority and environmental data make it unsuitable for managing environmental flows sustainably (Andreen, 2011).

References

Australian Bureau of Agricultural and Resource Economics (ABARE) (2010), "Assessing the Regional Impact of the Murray-Darling Basin Plan and the Australian Government's Water for the Future Programme in the Murray Darling Basin", ABARE-BRS Client Report for the Department of Sustainability, Environment, Water, Population and Communities, Canberra.

Allen Consulting Group (2007), "Saying Goodbye to Permanent Water Restrictions in Australia's Cities: Key Priorities for Achieving Water Security", *Report to Infrastructure Partnerships Australia*, February.

Andreen, W.L. (2011), "Water Law and the Search for Sustainability", in R.Q. Grafton and K. Hussey (eds.), *Water Resources Planning and Management*, Cambridge University Press, Cambridge.

Barry, C. (2011), "Brisbane Floods: Did the Dam Work?", *http://www.australiangeographic.com.au/journal/ brisbane-floods-did-the-dams-work.htm*.

Battalio, R.C., J.H. Kagel and C.A. Kogut (1991), "Experimental Confirmation of the Existence of a Giffen Good", *The American Economic Review*, Vol. 81, No. 4.

Bauer, C.J. (1998), *Against the Current: Privatization, Water Markets, and the State in Chile*, Kluwer Academic Publishers, Norwell.

Bauer, C.J. (1997), "Bringing Water Markets Down to Earth: The Political Economy of Water Rights in Chile, 1976-95", *World Development*, Vol. 25, No. 5.

Becker, N., D. Lavee and D. Katz (2010), "Desalination and Alternative Water-Shortage Mitigation Options in Israel: A Comparative Cost Analysis", *Journal of Water Resource and Protection*, Vol. 2.

Birol, E., P. Koundouri and Y. Kountouris (2010), "Assessing the Economic Viability of Alternative Water Resources in Water Scarce Regions: Combining Economic Valuation, Cost-benefit Analysis, and Discounting", *Ecological Economics*, Vol. 69, No. 4.

Biswas, A.K. (2008), "Integrated Water Resource Management: Is it Working?", *Water Resources Development*, Vol. 24, No. 1.

Bommelaer, O. et al. (2011), "Financing Water Resources Management in France (October 2011 Update)", Department of the Commissioner-General for Sustainable Development, Ministry of Ecology, Sustainable Development and Energy, Paris.

Boyd, J.W., L.A. Shabman and K. Stephenson (2007), "Trading as a U.S. Water Quality Management Tool: Prospects for a Market Alternative", *Ecological Economics of Sustainable Watershed Management*, Vol. 7.

Burnett, V. (2008), "Thirsty Barcelona Imports Drinking Water", *The New York Times*, *www.nytimes.com/ 2008/05/13/world/europe/13iht-spain.4.12853501.html*.

Business Insider (2011), "Citi's Top Economist Says the Water Market Will Soon Eclipse Oil", *www.businessinsider.com/willem-buiter-water-2011-7*.

Carey, J., D.L. Sunding and D. Zilberman (2002), "Transaction Costs and Trading Behaviour in an Immature Water Market", *Environment and Development Economics*, Vol. 7, No. 4.

Chong, H. and D.L. Sunding (2006), "Water Markets and Trading", *Annual Review of Environment and Resources*, Vol. 31.

Coase, R.H. (1960), "The Problem of Social Cost", *Journal of Law and Economics*, Vol. 3.

Colby, B. (1995), "Regulation, Imperfect Markets, and Transaction Costs: The Elusive Quest for Efficiency in Water Allocation", in D.W. Bromley (ed.), *The Handbook of Environmental Economics*, Blackwell Publishers, Cambridge.

Cowan, S. (1998), "Water Pollution and Abstraction and Economic Instruments", *Oxford Review of Economic Policy*, Vol. 14, No. 4.

Dalhuisen, J.M., H. de Grroot and P. Nijkamp (2000), "The Economics of Water: A Survey", *International Journal of Development Planning Literature*, Vol. 15, No. 1.

Daniell, K.A., I.S. Ribarova and N. Ferrand (2011), "Collaborative Flood and Drought Risk Management in the Upper Iskar Basin, Bulgaria", in R.Q. Grafton and K. Hussey (eds.), *Water Resources Planning and Management*, Cambridge University Press, Cambridge.

Dasgupta, P.S. and G.M. Heal (1979), *Economic Theory and Exhaustible Resources*, Cambridge University Press, Cambridge.

Dillon, P. (2011), "Water Security for Adelaide, South Australia", in R.Q. Grafton and K. Hussey (eds.), *Water Resources Planning and Management*, Cambridge University Press, Cambridge.

Dowd, B.M., D. Press and M.L. Huertos (2008), "Agricultural Non-Point Source Water Pollution Policy: The Case of California's Central Coast", *Agriculture, Ecosystems and Environment*, Vol. 128.

Duhon, M., H. McDonald and S. Kerr (2012), "Nitrogen Trading in Lake Taupo: An Analysis and Evaluation of an Innovative Water Management Policy," Motu Economic and Public Policy Research, Wellington.

Easter, K.W., M.W. Rosegrant and A. Dinar (eds.) (1998), *Markets for Water: Potential and Performance*, Kluwer Academic Publishers, Norwell.

Fisher, A. et al. (1995), "Alternatives for Managing Drought: A Comparative Cost Analysis", *Journal of Environmental Economics and Management*, Vol. 29.

Garcia-Valinas, M. (2006), "Analysing Rationing Policies: Droughts and its Affects on Urban Users' Welfare", *Applied Economics*, Vol. 38.

Gaudin, S. (2006), "Effect of Price Information on Residential Water Demand", *Applied Economics*, Vol. 38, No. 4.

Ghassemi, F. and I. White (2007), "Inter-Basin Water Transfer, Case Studies from Australia, United States, Canada, China and India", *International Hydrology Series*, Australian National University, Canberra.

Glachant, M. (2002), "The Political Economy of Water Effluent Charges in France: Why are Rates Kept Low?", *European Journal of Law and Economics*, Vol. 14.

Gleick, P.H. et al. (2002), "Globalization and the International Trade of Water" in P.H. Gleick et al., The World's Water, *The Biennial Report on Freshwater Resources 2002-03*, Island Press, Washington, DC.

Gomez-Lobo, A. (2002), "Making Water Affordable", in R.J. Brooke and S.M. Smith (eds.), *Contracting for Public Services: Output Based Aid and its Applications*, World Bank, Washington, DC.

Gomez-Lobo, A. and D. Contreras (2003), "Water Subsidy Policies: A Comparison of the Chilean and Colombian Schemes", *World Bank Economic Review*, Vol. 17, No. 3.

Grafton, R.Q. (2011), "Economic Instruments for Water Management", *background Paper presented to the OECD Working Party on Biodiversity, Water and Ecosystems*, Paris, 27-28 October, ENV/EPOC/WPBWE(2011)13.

Grafton, R.Q. (2010), "How to Increase the Cost-Effectiveness of Water Reform and Environmental Flows in the Murray-Darling Basin", *Agenda*, Vol. 17, No. 2.

Grafton, R.Q. et al. (2012), "A Comparative Assessment of Water Markets: Insights from the Murray-Darling Basin of Australia and the Western U.S.", *Water Policy*, Vol. 14, No. 2.

Grafton, R.Q. et al. (2011a), "An Integrated Assessment of Water Markets: A Cross-Country Comparison", *Review of Environmental Economics and Policy*, Vol. 5, No. 2.

Grafton, R.Q. et al. (2011b), "Determinants of Residential Water Consumption: Evidence and Analysis from a 10-country Household Survey", *Water Resources Research*, Vol. 47, No. W08537.

Grafton, R.Q. and Q. Jiang (2011), "Economics of Water Recovery in the Murray-Darling Basin", *The Australian Journal of Agricultural and Resource Economics*, Vol. 55.

Grafton, R.Q. and Q. Jiang (2010), "Economics of Drought, Water Diversions, Water Recovery and Climate Change in the Murray-Darling Basin", *CWEEP Research Paper*, No. 09-22, Australia National University.

Grafton, R.Q. and T. Kompas (2007), "Pricing Sydney Water", *The Australian Journal of Agricultural and Resource Economics*, Vol. 51, No. 3.

Grafton, R.Q. and M.B. Ward (2010), "Dynamically Efficient Urban Water Policy", *CWEEP Research Paper*, No. 10-13, Australia National University.

Grafton, R.Q. and M.B. Ward (2008), "Prices Versus Rationing: Marshallian Surplus and Mandatory Water Restrictions", *Economic Record*, Vol. 84, Issue Supplement s1.

Gruen, G.E. (2007), "Turkish Water Exports: A Model for Regional Cooperation", in H. Shuval and H. Dweik (eds.), *Water Resources in the Middle East, Israel-Palestinian Water Issues – From Conflict to Co-operation*, Springer, Berlin.

Haddad, B.M. (2000), "Rivers of Gold: Designing Markets to Allocate Water in California", *Island Press*, Washington, DC.

Hadfield, J. (2008), "Statement of Evidence of John Hadfield Before the Environment Court," New Zealand.

Hadjigeorgalis, E. and J. Lillywhite (2004), "The Impact of Institutional Constraints on the Limari River Valley Water Market", *Water Resources Research*, Vol. 40, No. W05501.

Hearne, R.R. and K.W. Easter (1997), "The Economic and Financial Gains from Water Markets in Chile", *Agricultural Economics*, 15.

Hidalgo, J. and H. Pena (2009), "Turning Water Stress into Water Management Success: Experiences in the Lerma-Chapala River Basin", in R. Lenton and M. Muller (eds.), *Integrated Water Resources Management in Practice: Better Water Management for Development*, Earthscan, London.

Hirji, R. and R. Davis (2009), "Environmental Flows in Water Resource Policies, Plans, and Projects: Case Studies", *World Bank Environment Department Papers*, Natural Resource Management Series, No. 117, The World Bank, Washington, DC.

Howard, G. and J. Bartram (2003), "Domestic Water Quantity, Service, Level and Health", WHO/SDE/WSH/03.02, World Health Organization (WHO), Geneva.

Howe, C.W. (1996), "Water Resources Planning in a Federation of States: Equity Versus Efficiency", *Natural Resources Journal*, Vol. 36.

Howe, C.W., D.R. Schurmeier and W. Douglas Shaw Jr. (1986), "Innovative Approaches to Water Allocation: The Potential for Water Markets", *Water Resources Research*, Vol. 22, No. 4.

Howitt, R.E. (1994), "Empirical Analysis of Water Market Institutions: The 1991 California Water Market", *Resource and Energy Economics*, Vol. 16, No. 4.

Hughes, N., A. Hafi and T. Goesch (2009), "Urban Water Management: Optimal Price and Investment Policy under Climate Variability", *Australian Journal of Agricultural and Resource Economics*, Vol. 53, No. 2.

International Conference of Water and the Environment (ICWE) (1992), "The Dublin Statement on Water and Sustainable Development", *www.wmo.int/pages/prog/hwrp/documents/english/icwedece.html*.

Kerr, S., M. Claridge and D. Milicich (1998), "Devolution and the New Zealand Resource Management Act," 98-7, New Zealand Treasury, Wellington, New Zealand.

Kerr, S., H. McDonald and K. Rutherford (2012), "Nutrient Trading in Lake Rotorua: A Policy Prototype,"*Motu Working Paper*, 12-10, Motu Economic and Public Policy Research, Wellington, New Zealand, *www.motu.org.nz/publications/detail/nutrient_trading_in_lake_rotorua_a_policy_prototype*.

Kraemer, R.A., E. Kampa and E. Interwies (2003), *The Role of Tradable Permits in Water Pollution Control*, Ecologic, Institute for International and European Environmental Policy, *http://ecologic.eu/1017*.

Lenton, R. and M. Muller (eds.) (2009), "Integrated Water Resources Management in Practice: Better Water Management for Development", *Earthscan Publications*, London.

Loux, J, (2011), "Collaboration and Stakeholder Engagement", in R.Q. Grafton and K. Hussey (eds.), *Water Resources Planning and Management*, Cambridge University Press, Cambridge.

MacDonald, D.H. et al. (2011), "Valuing a Multistate River: The Case of the River Murray", *The Australian Journal of Agricultural and Resource Economics*, Vol. 55.

Mansur, E.T. and S.M. Olmstead (2007), "The Value of Scarce Water: Measuring the Inefficiency of Municipal Regulations", National Bureau of Economic Research (NBER), *Working Paper*, 13513, Cambridge, United States.

Murray-Darling Basin Authority (MDBA) (2011), "Guide to the Proposed Basin Plan", *http://download.mdba. gov.au/Guide_to_the_Basin_Plan_Volume_1_web.pdf*.

Milliman, J.W. (1959), "Water Law and Private Decision-Making: A Critique", *Journal of Law and Economics*, Vol. 2.

Muller, M. (2009), "Attempting to Do it All: How a New South Africa has Harnessed Water to Address its Development Challenges", in R. Lenton and M. Muller (eds.), *Integrated Water Resources Management in Practice: Better Water Management for Development*, Earthscan, London.

National Water Commission (2010), *The Impacts of Water Trading in the Southern Murray-Darling Basin: An Economic, Social and Environmental Assessment*, National Water Commission, Canberra.

OECD (2010a), *Taxation, Innovation and the Environment*, OECD Publishing, *http://dx.doi.org/10.1787/ 9789264087637-en*.

OECD (2010b), *Pricing Water Resources and Water and Sanitation Services*, OECD Publishing, *http:// dx.doi.org/10.1787/9789264083608-en*.

OECD (2009), *Managing Water for All: An OECD Perspective on Pricing and Financing*, OECD Publishing, *http:/ /dx.doi.org/10.1787/9789264083608-en*.

OECD (2008), *OECD Environmental Outlook to 2030*, OECD Publishing, *http://dx.doi.org/10.1787/ 9789264040519-en*

OECD (2007a), *OECD Environmental Performance Reviews: Denmark 2007*, OECD Publishing, *http:// dx.doi.org/10.1787/9789264039582-en*.

OECD (2007b), *Instrument Mixes for Environmental Policy*, OECD Publishing, Paris, *http://dx.doi.org/10.1787/ 9789264018419-en*.

OECD (2006), *The Political Economy of Environmentally Related Taxes*, OECD Publishing, *http://dx.doi.org/ 10.1787/9789264025530-en*.

OECD (2003), *Social Issues in the Provision and Pricing of Water Services*, OECD Publishing, *http://dx.doi.org/ 10.1787/9789264099890-en*.

OECD (1999), *The Price of Water: Trends in OECD Countries*, OECD Publishing, *http://dx.doi.org/10.1787/ 9789264173996-en*.

Olmstead, S.M. (2010), "The Economics of Water Quality", *Review of Environmental Economics and Policy*, Vol. 4, No. 1.

Pahl-Wostl, C., P. Jeffrey and J. Sendzimir (2011), "Adaptive and Integrated Management of Water Resources", in R.Q. Grafton and K. Hussey (eds.), *Water Resources Planning and Management*, Cambridge University Press, Cambridge.

Peterson, D. et al. (2004), "Modelling Water Trade in the Southern Murray-Darling Basin", *Productivity Commission Staff Working Paper*, Productivity Commission, Melbourne.

Peszko, G. and P. Lenain (2001), "Encouraging Environmentally Sustainable Growth in Poland", *OECD Economics Department Working Papers*, No. 299, OECD Publishing, *http://dx.doi.org/10.1787/148456178784*.

Pittock, J. (2011), "Damned if You Do…", *The Global Water Forum*, Australian National University, Canberra, *www.globalwaterforum.org/2011/10/01/damned-if-you-do*.

Pittock, J. and C.M. Finlayson (2011), "Freshwater Ecosystem Conservation: Principles Versus Policy", in D. Connell and R.Q. Grafton (eds.), *Basin Futures: Reform in the Murray-Darling Basin*, Australia National University E-Press, Canberra.

Productivity Commission (2011), "Australia's Urban Water Sector", *Productivity Commission Draft Report*, Canberra.

Productivity Commission (2010), "Market Mechanisms for Recovering Water in the Murray-Darling Basin", Commissioned Study for the Australian Government, *www.pc.gov.au/projects/study/water-recovery*.

Qureshi, M.E. et al. (2010), "Environmental Water Incentive Policy and Return Flows", *Water Resources Research*, Vol. 46, No. W04517.

Rende, M. (2007), "Water Transfer from Turkey to Water Stressed Countries in the Middle East", in H. Shuval and H. Dweik (eds.), *Water Resources in the Middle East, Israel-Palestinian Water Issues – From Conflict to Co-operation*, Springer, Berlin.

Renwick, M.E. and R.D. Green (2000), "Do Residential Water Demand Side Management Policies Measure Up? An Analysis of Eight California Water Agencies", *Journal of Environmental Economics and Management*, Vol. 40, No. 1.

Rive, V. (2012), "Nutrient Trading in Lake Rotorua: Design, Implementation and Enforcement – Legal Issues," Vernon Rive Environmental Law, *www.motu.org.nz/publications/detail/nutrient_trading_in_lake_rotorua_design_implementation_and_enforcement_-_le*.

Rogers, P., R. De Silva and R. Bhatia (2002), "Water is an Economic Good: How to Use Prices to Promote Equity, Efficiency, and Sustainability", *Water Policy*, Vol. 4.

Rutherford, K., Ch. Palliser and S. Wadhwa (2011), "Prediction of Nitrogen Loads to Lake Rotorua Using the ROTAN Model", *Report Prepared for the Bay of Plenty Regional Council*, National Institute of Water and Atmospheric Research, Hamilton.

Saliba, B.C. and D.B. Bush (1987), "Water Markets in Theory and Practice: Market Transfers, Water Values, and Public Policy", *Westview Press*, Boulder, Colorado, United States.

Selman, M. et al. (2009), "Water Quality Trading Programs: An International Overview", *WRI Issue Brief: Water Quality Trading No. 1*, World Resources Institute, Washington, DC. *http://pdf.wri.org/water_trading_quality_programs_international_overview.pdf*.

Shaw, W.D. (2005), *Water Resource Economics and Policy*, Edward Elgar, Cheltenham.

Shortle, J.S. (2012), "Water Quality Trading in Agriculture," COM/TAD/CA/ENV/EPOC(2010)19/FINAL, OECD, Paris, France, *www.oecd.org/dataoecd/5/1/49849817.pdf*.

Shortle, J.S. and R.D. Horan (2001), "The Economics of Non-Point Pollution Control", *Journal of Economic Surveys*, Vol. 15, No. 3.

Sibly, H. (2006a), "Efficient Urban Water Pricing", *The Australian Economic Review*, Vol. 39, No. 2.

Sibly, H. (2006b), "Urban Water Pricing", *Agenda*, Vol. 13, No. 1.

Sibly, H. and R. Tooth (2008), "Bringing Competition to Urban Water Supply", *Australian Journal of Agricultural and Resource Economics*, Vol. 52, No. 3.

Stavins, R.N. (2003), "Experience with Market-Based Environmental Policy Instruments", in K.G. Maler and J.R. Vincent (eds.), *Handbook of Environmental Economics*, Vol. 1, Elsevier.

Tietenberg, T.H. (1990), "Economic Instruments for Environmental Regulation", *Oxford Review of Economic Policy*, Vol. 6, No. 1.

United States Environmental Protection Agency (US EPA) (2001), "The National Costs of the Total Maximum Daily Load Programme", *Draft Report*, Office of Water, U.S. EPA, Washington, DC.

Van Houtven, G., J. Powers and S.K. Pattanayak (2007), "Valuing Water Quality Improvements in the United States Using Meta-Analysis: Is the Glass Half-Full or Half-Empty for National Policy Analysis?", *Resource and Energy Economics*, Vol. 29.

Vant, W.N. (2008), "Statement of Evidence of William Nisbet Vant", before the Environment Court, January 2008, New Zealand.

Veblen, T. (1899), *The Theory of the Leisure Class; an Economic Study of Institutions*, MacMillan Company, New York.

Wachtel, H.M. (2007), "Water Conflicts and International Water Markets", in H. Shuval and H. Dweik (eds.), *Water Resources in the Middle East, Israel-Palestinian Water Issues – From Conflict to Co-operation*, Springer, Berlin.

Wentworth Group of Concerned Scientists (2010), "Sustainable Diversions in the Murray-Darling Basin: An Analysis of the Options for Achieving a Sustainable Diversion Limit in the Murray-Darling Basin", *www.wentworthgroup.org/uploads/Sustainable%20Diversions%20in%20the%20Murray-Darling%20Basin.pdf*.

Wheeler, S. et al. (2011), "Incorporating Temporary Trade with the Buy-Back of Water Entitlements in Australia", *Centre for Water Economics, Environment and Policy Research (CWEEP) Research Paper 11-01*, Australia National University.

Wheeler, S. et al. (2010), "Estimating South Murray Darling Basin Irrigators' Willingness to Sell Water Entitlements to the Federal Government", *Centre for Regulation and Market Analysis (CRMA) Working Paper*, University of South Australia.

Wilson, M.A. and S.R. Carpenter (1999), "Economic Valuation of Freshwater Ecosystem Services in the United States", *Ecological Applications*, Vol. 9, No. 3.

Wittwer, G. (2010), "The Economic Impacts of a New Dam in South-East Queensland", *Australian Economic Review*, Vol. 42, No. 1.

Woo, C. (1994), "Managing Water Supply Shortage: Interruption Versus Pricing", *Journal of Public Economics*, Vol. 54.

World Commission on Dams (WCD) (2000), *Dams and Development: A New Framework for Decision Making*, Earthscan Publications, London.

Worthington, A.C. and M. Hoffman (2008), "An Empirical Survey of Residential Water Consumption", *Journal of Economic Surveys*, Vol. 22, No. 5.

Young, J. (2007), "Statement of Evidence of Justine Young", before the Environment Court, January 2007, New Zealand.

Young, J., G. Kaine and Environment Waikato (2010), *Application of the Policy Choice Framework to Lake Taupo Catchment*, 20 ed., Hamilton: Environment Waikato.

Zetland, D. (2011), "The End of Abundance: Economic Solutions to Water Scarcity", *Aguanomics Press*, Amsterdam.

Chapter 4

Policy coherence toward water security

Water security should be pursued taking account of complex links with economic and sectoral policies. Setting acceptable levels of water risks among stakeholders should be the result of well-informed trade-offs between water security and other policy objectives. Meeting the coherence challenge also requires a coherent approach between water and other (sectoral, environmental) policies.

In view of the gloomy outlook regarding water stress and water pollution, and the growing uncertainties regarding floods and droughts, governments need to speed up efforts to enhancing overall efficiency and effectiveness in water management to alleviate growing water security concerns. As explained in previous Chapters, this entails better risk management and better water policies. But water security should also be pursued taking account of complex links with economic and sectoral policies.

Allocating water risks between residential, agricultural, industrial and environmental uses raises a significant political economy question. As explained in Chapters 1 and 2, setting acceptable levels of water risks among stakeholders is one of the most challenging and controversial tasks in the risk management process. It should be the result of well-informed trade-offs between water security and other (sectoral, environmental) policy *objectives*.

Meeting the coherence challenge also requires a coherent approach between water policies and other (sectoral, environmental) policies (in part, following OECD, 2012a).

In particular, the nexus between water, energy, food and the environment presents significant challenges for water security, and has been attracting increasing policy attention in recent years. Increasing the coherence of policies (policy objectives and policy instruments) across these areas is essential if governments wish to meet the range of policy goals while not undermining water security objectives.

The linkages between water and energy are important and pervasive. The importance of water in energy production and use is matched by the importance of energy in water. As countries confront water resource constraints, their arsenal of policy options has typically included energy-intensive solutions such as long haul transfer and desalination. The corollary is also true: many countries address energy constraints with water-intensive options such as steam-cycle power plants or biofuels. However, this approach, whereby water planners assume they have all the energy they need and energy planners assume they have all the water they need, is not likely to work effectively in the future. Countries that deploy incoherent water and energy policies might find themselves with severe scarcity of one resource or the other, or both.

Similarly, water and agriculture are inextricably linked, not least because agriculture accounts for around 70% of water withdrawals globally. Support provided to lower the costs of water supplied to agriculture, for example, by not reflecting the scarcity value of water, can undermine efforts to achieve water security objectives. Agricultural support policies linked to production can also jeopardise water security through providing incentives to intensify and extend production more than would be the case in the absence of this form of support. But isolating and quantifying the overall economic efficiency and environmental effectiveness of agricultural support on water is difficult and further analysis on causation is needed.

Policies across water, energy, agriculture and environment are often formulated without sufficient consideration of their inter-relationship or their unintended consequences. The silo nature of many governments' approaches to policy development in the different areas is the key contributor to this incoherence. This translates into differences in temporal scales between energy, agricultural and water policy objectives (e.g. forward-looking water plans are often on the 50-60 year horizon, whereas energy plans are up to 20-30 years ahead, and agricultural planning is generally within a much shorter time horizon).

Success in achieving greater coherence between energy, agriculture and water policies will ultimately depend on removing policy inconsistencies, especially where energy and agricultural support policies conflict with water security goals. More coherent policy approaches are slowly beginning to take shape in a growing number of OECD countries. For example, lowering overall agricultural support and shifting from direct production and input agricultural support to decoupled payments over the past 20 years in many OECD countries has, in part, led to improvements in water resource use efficiency and helped to lower water pollution pressure from agricultural activities. But much more needs to be done in both OECD and non-OECD countries.

Options to enhance policy coherence include exploiting win-wins (such as taking steps to increase both water and energy efficiency) and assessing and managing trade-offs between (sometimes conflicting) policy objectives.

Spillover effects of sectoral and environmental policies on water security

By creating incentives towards meeting their own objectives, sectoral (e.g. agricultural, energy) and environmental (e.g. climate, biodiversity) policies may have significant spillover to water security. The links between water and other related security objectives – food, energy, climate, biodiversity – are not routinely addressed or fully understood. Yet uncoordinated policy aimed at security in one area may result in less security in another: less water security as the cost of greater energy security through biofuel production, for example (Zeitoun, 2011).

Complexity arises from the need to consider the direct and indirect impacts of sectoral policies on water security. The same sectors (e.g. agriculture, energy) that impact on water also impact on other components of the environment (e.g. climate, nature). Moreover, within a sector, the objectives of environmental protection and improving water management sometimes conflict with each other (e.g. subsidies to fast-growing forest plantations aimed at carbon sequestration are sometimes at the detriment of old growth natural forests that better regulate water flows).

When they last met at the OECD in 2010, Ministers of Agriculture from OECD member countries and key emerging economies recognised that an integrated approach to **food security** is needed involving a mix of domestic production, international trade, stocks, safety nets for the poor, and other measures reflecting levels of development and resource endowment, while poverty alleviation and economic development are essential to achieve a sustainable solution to global food insecurity and hunger in the longer term. If people are hungry today, it is because they cannot afford to buy food, not because there is not enough available.

The Millennium Development Goals of halving the share of the global population suffering from hunger in 2015 compared with 1990 will not be met. Increasing productivity, establishing (and enforcing) well-defined land property rights and ensuring that well-functioning agricultural markets provide the right signals are the three priority areas where coherent action is required if the additional one billion tonnes of cereals and 200 million tonnes of meat that would need to be produced annually between now and 2050 to feed everyone is to be produced without over-exploiting scarce natural resources or further damaging the environment (including water) (OECD, 2011a).

Food security impacts on water security through agricultural policy distorting production and trade of agricultural commodities, thereby distorting the domestic and global demand for water.

In the agricultural sector, irrigation has to some extent helped with climatic risk management, thus reducing pressures on governments to compensate for flood damage downstream or for crop losses as a result of periodic droughts. However, below-cost pricing is prevalent for publicly-funded irrigation systems. It is, in the main, national treasuries that have financed dams, reservoirs and delivery networks, as well as a large part of the cost of installing local and farm infrastructure. Governments generally attempt to recover some of these costs through user charges, but revenues are rarely enough to cover even operation and maintenance costs.

The economic distortions caused by the often enormous under-pricing of water used in agriculture have been compounded in many instances by agricultural policies, particularly those linked to the production of particular commodities. Such linked support draws resources, including water, into the activity being supported, thereby driving up both the price of water to other users and the volume of agricultural subsidies. As a rule, farmers have free access to (or are charged only a nominal fee for) water that they pump themselves. And several countries continue to offer preferential tariffs for electricity used to pump water for irrigation.

There are conflicting views about whether trade in virtual water can lead to overall savings in global water resources (Lenzen et al., 2012). Countries are experiencing vastly different degrees of water scarcity. There is indirect virtual water use throughout the supply chains underlying all traded goods. When adjusting water volumes for water scarcity and when indirect virtual water is appraised the Heckscher-Ohlin Theorem can be validated.[1] In other words, trade liberalisation tends to reduce water use in water scarce regions and increase water use in water abundant regions.

However, the global impact of agricultural trade liberalisation and policy reform on water systems is likely to be limited. Research suggests that the impact of hypothetical Doha-like liberalisation of agricultural trade on water use would be a change in regional water use of less than 10%, even if agricultural tariffs are reduced by 75% (Berrittella et al., 2007). Patterns are non-linear: water use may go up for partial liberalisation, and down for more complete liberalisation. This is because different crops respond differently to tariff reductions, but also because trade and competition matter too.

Moreover, there has already been a major reduction in overall agricultural support in OECD countries over the past 20 years, including production and input related support, limiting impacts of further liberalisation. Other drivers are having a much greater impact on global water systems that agricultural support, notably increasing agricultural production and rising trend of world commodity prices. For example, there is a strong correlation between increases in world dairy prices, rising cow numbers and increasing nutrient water pollution in New Zealand.

Energy policy makers are facing the daunting challenge of achieving energy security, environmental protection and economic efficiency (the three Es). The need to increase **energy security** was the main objective underpinning the establishment of the International Energy Agency (IEA). According to the IEA, energy security can be described as "the uninterrupted physical availability at a price which is affordable, while respecting environment concerns". IEA member countries co-operate to increase their collective energy security through diversification of their energy sources and improved energy efficiency.

Energy security impacts on water security through increasing the water needs and water pollution linked to increased energy supply or further reliance on renewable energies, such as hydropower and biofuels.

Oil security remains a cornerstone of the IEA.[2] At the same time, the IEA is progressively taking a more comprehensive approach to the security of supplies, including natural gas supplies and power generation.

A universal phase-out of all fossil fuel consumption subsidies by 2020 – ambitious though it may be as an objective – would cut global primary energy demand by 5%, compared with a baseline in which subsidies remain unchanged (IEA, 2010). Reducing reliance on fossil fuels would also impact on the competition between food and biofuels for water (water for energy), which is directly related to the demand (and cost) of fossil fuels.

Support to agricultural feedstocks to produce biofuels and bioenergy has been increasing in recent years. Such support can have significant impacts on water quality and availability. The water quality impacts may be caused by the use of agrochemicals in intensive bioenergy feedstock production systems (OECD, 2012b). The impact on water balances remains unclear. It is largely an empirical question and needs to be assessed in a way that compares the effects of alternative uses of resources (OECD, 2010a). Research suggests that the quantity of water needed to produce each unit of energy from second generation biofuel feedstocks (e.g. lignocellulosic harvest residues and forestry) is much lower than the water required to produce ethanol from first generation feedstocks (such as from cereals, oilseeds and sugar crops), although this can vary according to the location and practices adopted to produce these different feedstocks.

Renewable energy sources will have to play a central role in moving the world onto a more secure, reliable and sustainable energy path. The potential is unquestionably large, but how quickly their contribution to meeting the world's energy needs grows hinges critically on the strength of government support to stimulate technological advances and make renewables cost competitive with other energy sources. Government support for renewables can, in principle, be justified by the long-term economic, energy security and environmental benefits they can bring, though it is essential that support mechanisms are cost-effective (IEA, 2010). Nearly all OECD countries have introduced renewable energy targets with a view to curb greenhouse gas emissions. However, such targets have proved to be a very expensive method of reducing greenhouse gas emissions compared with other abatement options, costing several times as much as the carbon taxes that have been introduced and well above the price in carbon cap-and-trade schemes. Apart from lowering carbon emissions, the expansion of renewable energy has been pursued for other reasons, such as reducing air pollution, strengthening energy security, raising employment levels, and increasing innovations. There is little or no evidence that such non-greenhouse gas related benefits justify the "excess" abatement costs or that special high support to renewables is the most efficient way to achieve such objectives.

The greatest scope for increasing the use of renewables in absolute terms lies in the power sector. The share of renewables in global electricity generation is expected to increase from 19% in 2008 to almost one-third (catching up with coal) by 2035 (IEA, 2010).

The increase is expected to come primarily from wind and hydropower, with hydropower remaining the most common form of renewable energy.

Since 1990, global hydropower generation has increased by 50%, with the highest absolute growth in China (OECD/IEA, 2010). Hydropower contributes to energy security and climate protection, being a renewable energy technology. When produced in storage schemes (e.g. storing water through dams),[3] it also brings water security benefits, through the supply of drinking water or water for irrigation and flood/drought risk management. In some cases, however, these benefits come at important social costs (i.e. displacement of people) and environmental costs (i.e. changes in flow and continuity of rivers). Brokered by the World Bank and the World Conservation Union (IUCN), a temporary World Commission on Dams (WCD) was established between 1998 and 2000 in response to the escalating local and international controversies over large dams. Its final report recommends that decisions on major infrastructure developments take place within a framework that recognises the rights of all stakeholders, and the risks that each stakeholder group is asked, or obliged to sustain (WCD, 2000). There is a need for cost benefit analysis prior to any project of building a new dam or retrofitting old ones.

As a natural resource, water is obviously influenced by climatic factors. What comes immediately to mind when addressing the interface between **climate policy** and water resources are water quantity issues (floods and droughts). But there are also water quality implications. For example, reducing the use of nitrogen fertilisers to curb greenhouse gas (nitrous oxide) emissions also reduces nitrate pollution. Water quantity and water quality are both part of the equation.

The benefits of climate change mitigation are long-term. Even if strong action was taken today, there would be no discernible effect (identifiable benefit) on rates of warming (and rainfall distribution) for considerable periods of time. Thus assessing the spillover effects of mitigation policy on water security entails looking at the ancillary benefits of mitigation. For instance, using hydropower to reduce carbon dioxide emissions can contribute to flood control through construction of dams and water reservoirs. It also entails looking at the ancillary costs. For example, hydropower dams may impose fish population relocation or could cause significant methane emissions (e.g. when vegetation covered by the dam decomposes).

There is concern that adaptation to climate change may greatly increase the costs of providing water infrastructure (Hughes et al., 2010). The water infrastructure design shall evolve, for example, to avoid disruption of biological sewage treatment (which does not operate well under high temperatures) or to reduce siltation in dams (due to increased soil erosion). Existing capital stock may have to be replaced quicker than expected, such as water supply reservoirs and flood control dykes, or displaced in the case of low-lying and coastal areas threatened by flooding and rising sea level. In regions becoming dryer with climate change, the scope for increasing usage of natural water supplies is reduced, and alternative supplies (desalination, water re-use) are costly.

As is the case for mitigation, information on the ancillary costs and benefits of adaptation policy would certainly contribute to better integrate adaptation concerns into water security planning. For example, the ancillary benefits of adaptation on flood risk management include restoration of natural habitats (in floodplains).

Understanding the effects of mitigation and adaptation policies on water security, and the interactions between them, is essential. For example, it may prove more cost-effective to support the creation of wetlands (in which bacteria convert nitrate to nitrogen released to the atmosphere) than to encourage organic farming or afforestation of farmland (to reduce the level of fertilisation).

Climate policy appears to have significant spillover to other policy areas that affect water security. This includes, *inter alia,* sectors as diverse as energy, transport, agriculture, forestry, fisheries and tourism. Information on such indirect water impacts of climate policy would certainly contribute to better integration of water security concerns by such sectors.

For example, in New Zealand, the carbon emission trading scheme (ETS) led to convert pastoral land to forestry, which also contributes to reduce nitrogen leaching into water. Innovative agreements have been made between farmers and major greenhouse gas (GHG) emitters (through a Trust) where the latter receive ETS credits in exchange of the former converting pastoral land to forestry, which also contributes to reduce nitrogen leaching into water (OECD, 2011b). This is occurring in Lake Taupo, the largest lake in the country, in danger of degradation due to agricultural effluent. A nitrogen cap-and-trade system was put in place for farmers around the lake. Instead of trading their nitrogen pollution rights, farmers can opt for permanent reductions in nitrogen, for which they are financially compensated through the Trust. In turn the Trust is financed by major GHG emitters, through purchases of ETS credits. Farmers are paid for the reduction in nitrogen emissions, at the same time as they receive income from forestry credits.

Healthy ecosystems underpin water security. Most notably, nature plays a very important role in regulating water flows. Healthy ecosystems reduce runoff (and therefore flood levels of the streams flowing from preserved areas) and improve water infiltration into the soil (helping to replenish the ground water). Nature also plays a role of purification of water resources, thus contributing to better water quality. For example, almost 1 million urban dwellers rely on natural wetlands for wastewater retention and purification services (WWAP, 2009). Healthy ecosystems can also enhance food security and climate security with spillover effects on water security. For example, healthy ecosystems help produce more food from each unit of agricultural land and improve resilience to climate change (Boelee, 2011).

A key step in addressing water risks is to understand ecosystems better and to seek to optimise the range of goods and services these ecosystems can provide to enhance water security. Greater coherence could be sought between water security and ecosystem protection objectives. For example, the World Wildlife Fund for Nature (WWF) is working toward the protection and management of 250 million hectares of representative wetlands worldwide.

To the extent that pressures on ecosystems increase water risks, **nature protection policy** can enhance water security. For example, to address flood risks technical engineering approaches of making/reinforcing dykes in lowland/downstream areas or lower river deltas are often seen as the most cost-effective option to protect densely populated and economically important areas. However, investments in land-use changes and floodplain restoration can be justified economically in the long run if, besides the expected value of the damage avoided, the additional non-priced socioeconomic benefits associated with these measures are taken into account (Brouwer and Van Ek, 2004). The net welfare gain would then also include improving river accessibility for recreational reasons and conserving high levels of biodiversity.

An interesting development of nature protection policy is the rapid increase in payments for ecosystem services (PES) over the past decade. As a voluntary, flexible, incentive-based and site-specific instrument, PES can provide potentially large gains in cost effectiveness compared to indirect payments or other regulatory approaches used for water security objectives (OECD, 2010b). PES is a mechanism under which the user or beneficiary of an ecosystem service makes a direct payment to an individual or community whose land use decisions have an impact on the ecosystem service provision (e.g. reducing water risks). The payments compensate individuals, such as farmers or foresters, for the additional costs of biodiversity and ecosystem service conservation and sustainable use, over and above that which is required by any existing regulations.

A criticism, however, is that PES fail to realise their potential cost-effectiveness gains. This is because PES programmes often make fixed uniform payments on a per hectare basis while biodiversity and ecosystem benefits tend to vary from one location to another. Moreover, individual holders of land-use rights are likely to have different opportunity costs of ecosystem service provision. PES programmes should be designed to take these differences into account.

For a PES programme to produce clear and effective incentives any conflicting market distortions, such as environmentally-harmful subsidies, should be removed. For example, policy intervention to further enhance the water security services unique to forests should not imply giving more subsidies to forest owners (to improve forest management) or to farmers (to convert farmland to forest). That would run the risk of repeating in the forestry sector the mistakes that policy reforms are now seeking to address in the agricultural sector. The reform of agricultural policy underway in OECD countries has in itself important implications for farmland conversion into forests: where price support to commodities is reduced, there is less incentive to expand agricultural production on marginal land. Instead of seeking compensation for any foregone revenues (from timber sales or from farming), any forestry payments should reward the provision of well-targeted and otherwise unremunerated water security services.

Currently, there are few examples where government has coordinated negotiations between potential beneficiaries and providers of ecosystem services but not directly funded the services. As with negative externalities, positive externalities are of public interest only where transaction costs are too high for those with direct benefits to coordinate with providers. Payments for ecosystem services between private actors that do not require government coordination are just normal market activity.

Effects of non-water environmental markets on water security: Some empirical evidence

Ecosystem service values are often addressed (and even modelled) as though they were independent. In reality the marginal value of an ecosystem service changes when complementary or conflicting ecosystem services are regulated. An existing regulation can either reduce or increase the value and cost of regulating a second ecosystem service. One example of this is the interaction between land-related climate change mitigation and water quality. Others would be links between water quality and quantity, climate change and water quantity, and any of these and biodiversity values. These interactions occur for all forms of regulation but are particularly visible with market-based instruments and especially environmental markets where allowance prices are visible and the cost of regulation and its distribution depends not only on abatement costs but also on the value and initial distribution of allowances.

This section considers interactions among environmental markets, with an empirical focus on two markets, for water quality and greenhouse gas emissions from land use. This helps identify how the interaction of externalities and markets can lead to unexpected water risks, but also opportunities to reduce water risk and ease the path for regulation.

In New Zealand, the Lake Taupo water quality market has vividly illustrated the potential for positive interaction between land-related climate change regulation and water quality regulation. Nearly all trades to date have involved some land conversion into forestry (Duhon et al., 2012). These farmers have not only sold nitrogen allowances, but have also sold carbon credits through New Zealand's emissions trading system (Mighty River Power, 2010).

In the Lake Rotorua catchment, Yeo et al.(2012) have modelled the interactions between these markets for the planned Lake Rotorua catchment nutrient trading system, the existing forestry component of the New Zealand Emission Trading System (ETS) (Karpas and Kerr, 2011) and the potential regulation of agricultural greenhouse gas emissions in New Zealand (Kerr and Sweet, 2008).

They find that greenhouse gas emissions trading alone can lead to large gains in water quality, while water quality trading has even larger impacts on greenhouse gas emissions (in this case where the nitrogen cap is very stringent) (Table 4.1). For sheep/beef farmers, the loss of farm profits as farmers de-intensify and in some cases convert to forestry is larger under the combination of both regulations because their profitability in sheep/beef production becomes so low relative to alternative uses. In contrast, for dairy farmers, the combination of two regulations makes it easier to stay in dairy farming than under water quality regulation alone. This is because the strong mitigation response by sheep/beef farmers to the combined regulation reduces their demand for nitrogen allowances, lowers the price of nitrogen allowances in the catchment, and makes it more profitable for dairy farmers to pay for nitrogen and continue to farm.

Table 4.1. **Effects and costs of combined greenhouse gas and nitrogen policies, Lake Rotorua**

			Sheep/beef farms		Dairy farms	
	N leaching	Net GHG emissions	Abatement cost (loss of profit from farming)	Economic profit (including permit cost and revenue)	Abatement cost (loss of profit from farming)	Economic profit (including permit cost and revenue)
	(tonnes/year)		(USD/ha/year)			
No regulation	506	137 133	–	480	–	1 369
GHG only	392	70 239	43	423	42	1 041
N only	134	-34 415	126	152	937	92
Both N and GHG	134	-75 663	409	246	448	245

Note: Scenario with no free allocation of nitrogen allowances. N: nitrogen; GHG: greenhouse gas.
Source: Derived from Yeo et al.(2012).

Another interesting impact of the combined regulations is that both sheep/beef and dairy farmers are better off with the GHG (emissions trading) regulation as well as the nitrogen cap if they are required to purchase all their allowances (and able to sell carbon credits) (Table 4.1). For sheep/beef, the benefit comes from carbon credit revenue; for dairy, it is because of the fall in the cost of the nitrogen allowances they purchase.

A contrasting case, where the two environmental markets could come into conflict arises in the Manawatu catchment (Manawatu-Wanganui region, New Zealand) where the emissions trading policy can induce land conversion into maize which is associated with high nitrogen losses (Daigneault et al., 2012). They also find that if an emissions trading system is already operating, the addition of a nutrient trading system could lead to real environmental gains at relatively low cost. In contrast, if the water quality regulation already exists (with a low level of stringency) adding the GHG regulation provides little gain at high cost. Clearly the interactions are sensitive to local conditions.

Thus the marginal environmental value from additional regulation is sensitive to the existing regulation for other related services. This is true of environmental markets but also of other market-based instruments. When several ecosystem service markets or payment/tax systems interact it is critical to take account of the interactions between them. Many efforts to value ecosystem services in order to provide payments ignore this. A payment for one ecosystem service (e.g. greenhouse gas mitigation) reduces the marginal value of complementary ecosystem services (e.g. water quality).

A framework for managing trade-offs between policies

As we have seen, pursuing sectoral and environmental policy objectives may have significant spillover to water security. Water policies have an important role to play in the overall mix of policies to achieve water security objectives, but sectoral and economic policies often play the most important role. Appropriate co-ordination and coherence therefore needs to be embodied within this policy mix – both domestically and internationally.[4]

This co-ordination is most likely to be effective when due account is taken of water security objectives in the initial establishment of objectives in non-water policy areas. This implies that sectoral decision makers should systematically undertake both *ex ante* and *ex post* assessments of the water impacts of their activities. This can usually best be achieved through the use of evaluation tools, especially environmental impact assessments, regulatory impact assessments and cost-benefit analysis.

Subsidies for various economic purposes are pervasive, both in OECD countries and worldwide. Many subsidies distort prices and resource allocation decisions, altering the pattern of production and consumption within the domestic economy and across countries. As a result, subsidies can have unforeseen negative effects on water. For example, agricultural subsidies can lead to overuse of pesticides and fertilisers; coal subsidies can substitute natural gas for more water intensive energy source such as coal; fuel tax rebates and subsidies for road transport increase eutrophying depositions of nitrogen oxides (NO_X).[5]

Regular efforts should therefore be made to identify those subsidies whose removal (or reform) would benefit water security. A quick scan of these subsidies would likely be sufficient to understand the main effects that subsidy reform would have on the decisions of consumers and producers, as well as the key linkages between those decisions and water. This would also provide an initial ranking of subsidies in terms of their harmfulness to water security.

Using one subsidy to offset the negative environmental effects of another is likely to be both ineffective and inefficient. In most case, reforming both of these subsidies will be a better solution. For example, high levels of production-linked price support have traditionally been provided to the agriculture sector. This has encouraged overuse of chemical inputs, as well as expansion of farming onto land that is of relatively low value economically – but often of high value to protect water systems. In turn, this has led to efforts to address these negative environmental impacts via subsidies that are conditional on meeting certain environmental standards (cross-compliance). It will generally prove to be more efficient and effective to reform the original subsidy than to retain (and try to correct) the environmental problems it creates through cross-compliance requirements.

A major factor that can promote the reform of environmentally harmful subsidies is increased transparency. It should therefore be made clear to the public-at-large who is benefiting from existing subsidy programmes, and the conditions under which these subsidies are being provided.

Conclusions

Food security impacts on water security through agricultural policy distorting production and trade of agricultural commodities, thereby distorting the domestic and global demand for water. The economic distortions caused by the often enormous under-pricing of water (or the electricity to pump water) used in agriculture are compounded by agricultural policies, particularly those linked to the production of particular commodities. Such linked support draws water into the activity being supported, thereby driving up both the price of water to other users and the volume of agricultural subsidies. There is need to pursue efforts toward agricultural policy reform.

However, the impact of agricultural trade liberalisation and policy reform on regional water use (e.g. water-abundant countries exporting more water-intensive goods) is likely to be limited. This is because of two main reasons. First, there has already been a major reduction in overall agricultural support in OECD countries over the past 20 years, including production and input related support, limiting impacts of further liberalisation. And, most importantly, other drivers are having a much greater global impact on water risks that agricultural support, notably increasing agricultural production and rising trend of world commodity prices.

Energy security affects water security through increasing the water needs and impairing the water quality linked to increased energy supply and changes in the energy mix. For example, coal subsidies encourage energy consumption, which may increase water risks; coal subsidies can also substitute natural gas for coal, a more water intensive energy source.

Energy security also affects water security insofar as it promotes further reliance on renewable energies, such as hydropower and biofuels. When produced in storage schemes, the expansion of hydropower can bring water security benefits through increasing freshwater supply and improving flood/drought risk management. There is little or no evidence, however, that government support to hydropower is the most efficient way to achieve such objectives. Moreover, the benefits of hydropower may come at important social (e.g. displacement of people) and environmental (e.g. changes in flow and continuity of rivers) costs. There is a need for cost benefit analysis prior to any project of building a new hydropower dam or retrofitting old ones.

Support to agricultural feedstocks to produce biofuels and bioenergy has been increasing in recent years. Such support can have significant impacts on water quality and availability. The water quality impacts may be caused by the use of agrochemicals in intensive bioenergy feedstock production systems. The impact on water balances remains unclear, though. It will depend on the extent to which advanced biofuels – whose feedstock crops tend to be less water-intensive – penetrate markets but foremost on the location and practices adopted to produce these different feedstocks.

Understanding the effects of **climate mitigation and adaptation policies** on water security, and the interactions between them, is essential. For example, where the objective is to manage the risk of nitrate pollution of water, an adaptation policy to expand natural floodplains through supporting the creation of wetlands (in which bacteria convert nitrate to nitrogen released to the atmosphere) may prove more cost-effective than a mitigation policy to reduce nitrous oxide (N_2O) by encouraging organic farming (to reduce the level of fertilisation).

Moreover, climate policy appears to have significant spillover to other policy areas that affect water security. This includes *inter alia* sectors as diverse as energy, transport, agriculture, forestry, fisheries and tourism. Information on such indirect water security impacts of climate policy would certainly improve economic efficiency (e.g. avoiding farmers to be paid for the reduction in nitrogen emissions at the same time as they receive income to convert farmland to forest land, which also contributes to reduce nitrogen leaching into water) and social welfare (e.g. air quality co-benefits of mitigating carbon emissions improve human health and reduce eutrophying depositions on surface water).

To the extent that pressures on ecosystems increase water risks, **nature protection policy** can enhance water security. As a flexible, incentive-based and site-specific instrument, payments for ecosystem services (PES) can provide potentially large gains in cost effectiveness compared to indirect payments or other regulatory approaches to manage water risks. For this, the payments should only compensate holders of land-use rights (e.g. farmers or foresters) for the additional costs of ecosystem service provision, over and above legal requirements. They should not take the form of uniform payments on a per hectare basis, as is often the case, but take account of differences in ecosystem benefits and opportunity costs for holders of land-use rights.

Any conflicting market distortions should be removed. For example, policies to enhance the water security services unique to forests should not imply giving more subsidies to foresters to improve forest management. That would run the risk of repeating in the forestry sector the mistakes that policy reforms are now seeking to address in the agricultural sector. Instead of seeking compensation for any foregone revenues from timber sales, any forestry payments should reward the provision of well-targeted and otherwise unremunerated water security services.

Ecosystem service values are often addressed as though they were independent. In reality there are **interaction between ecosystem services.** When several policy instruments to promote ecosystem service interact it is critical to take account of the interactions between them. Many efforts to value ecosystem services in order to provide payments ignore this. A payment for one ecosystem service (e.g. greenhouse gas mitigation) reduces the marginal value of complementary ecosystem services (e.g. water quality).

Decoupling **subsidies** from the use of water resources, as well as from production and consumption activities that harm the water environment, can yield important benefits to water security. This approach is fundamentally more coherent than one which promotes economic goals in isolation of water security considerations.

Notes

1. The Heckscher-Ohlin (H-O) Theorem states that the water-abundant country A exports the water-intensive good, while the capital-abundant country B exports the capital-intensive good.

2. For example, each IEA member is required to hold oil stocks equivalent to at least 90 days of net imports.

3. As opposed to run-of-river schemes, which use the natural flow of a river.

4. In part, following OECD, 2008.

5. Eutrophication results from discharges of nitrogen and phosphorus to inland and coastal waters.

References

Berrittella, M. et al.; (2007), "The Impact of Trade Liberalisation on Water Use: A Computable General Equilibrium Analysis", *Working Paper*, FNU-142, Research Unit Sustainability and Global Change, Hamburg University and Centre for Marine and Atmospheric Science. *www.fnu.zmaw.de/fileadmin/fnu-files/publication/working-papers/cgewaterdohawp.pdf*.

Boelee, E. (2011), *Ecosystems for Water and Food Security*, UNEP Nairobi, International Water Management Institute (IWMI), Colombo.

Brouwer, R. and R. Van Ek (2004), "Integrated Ecological, Economic and Social Impact Assessment of Alternative Flood Protection Measures in the Netherlands", *Ecological Economics,* Vol. 50 (1-2).

Daigneault, A., S. Greenhalgh and O. Samarasinghe (2012), "Economic Impacts of GHG and Nutrient Reduction Policies in New Zealand: A Tale of Two Catchments,"*Paper prepared for the Agricultural and Applied Economics Association (AAEA) Organised Symposium* at the 56th Annual Australian Agricultural and Resource Economics Society Annual Meeting, Fremantle, Western Australia, 8-10 February 2010.

Duhon, M., H. McDonald and S. Kerr (2012), "Nitrogen Trading in Lake Taupo: An Analysis and Evaluation of an Innovative Water Management Policy," Motu Economic and Public Policy Research, H.G. Wellington et al. (2010), "The Costs of Adaptation to Climate Change for Water Infrastructure in OECD Countries", *Utilities Policy* (2010), *http://dx.doi.org/10.1016/j.jup.2010.03.002*.

Hughes, G. et al. (2010), "The Costs of Adaptation to Climate Change for Water Infrastructure in OECD Countries", *Utilities Policy* (2010), *http://dx.doi.org/10.1016/j.jup.2010.03.002*.

International Energy Agency (IEA) (2010), *World Energy Outlook 2010*, IEA, Paris.

Karpas, E. and S. Kerr (2011), "Preliminary Evidence on Responses to the New Zealand Forestry Emissions Trading System", *Motu Working Paper 11-09*, Motu Economic and Public Policy Research, Wellington, New Zealand, *www.motu.org.nz/publications/detail/preliminary_evidence_on_responses_to_the_new_zealand_forestry_emissions_tra*.

Kerr, S. and A. Sweet (2008), "Inclusion of Agriculture in a Domestic Emissions Trading Scheme: New Zealand's Experience to Date", *Farm Policy Journal*, 5:4.

Lenzen, M. et al.(2012), "The Role of Scarcity in Global Virtual Water Flows", Center for Development Research (ZEF), *ZEF-Discussion Papers on Development Policy*, No. 169, Bonn University, September.

Mighty River Power (2010), "Media Release: Carbon Transaction Supports Environmental Benefits to Lake Taupo," Mighty River Power Media Centre, *www.mightyriver.co.nz/Media-Centre/Latest-News/2010-Archive/Carbon-Transaction-Supports-Environmental-Benefits.aspx*.

OECD (2012a), *Meeting the Water Reform Challenge*, OECD Publishing, *http://dx.doi.org/10.1787/9789264170001-en*.

OECD (2012b), *Water Quality and Agriculture: Meeting the Policy Challenge*, OECD Studies on Water, OECD Publishing, *http://dx.doi.org/10.1787/9789264168060-en*.

OECD (2011a), *Food and Agriculture*, OECD Green Growth Studies, OECD Publishing, *http://dx.doi.org/10.1787/9789264107250-en*.

OECD (2011b), *OECD Economic Surveys: New Zealand 2011*, OECD Publishing, *http://dx.doi.org/10.1787/eco_surveys-nzl-2011-en*.

OECD (2010a), *Sustainable Management of Water Resources in Agriculture*, OECD Studies on Water, OECD Publishing, *http://dx.doi.org/10.1787/9789264083578-en*.

OECD (2010b), *Paying for Biodiversity: Enhancing the Cost-Effectiveness of Payments for Ecosystem Services*, OECD Publishing, *http://dx.doi.org/10.1787/9789264090279-en*.

OECD (2008), "An OECD Framework for Effective and Efficient Environmental Policies", *Paper presented at the Meeting of the Environment Policy Committee* (EPOC) at Ministerial Level, 28-29 April 2008, ENV/EPOC(2008)7/FINAL.

OECD/IEA (2010), "Renewable Energy Essentials: Hydropower", OECD/IEA, Paris, *www.iea.org/papers/2010/Hydropower_Essentials.pdf*.

World Commission on Dams (WCD) (2000), "Dams and Development: A New Framework for Decision-Making", the *Report of the World Commission on Dams*, November 2000, Earthscan Publications Ltd., London and Sterling, VA.

World Water Assessment Programme (WWAP) (2009), *The United Nations World Water Development Report 3: Water in a Changing World*, UNESCO, Paris and Earthscan, London.

Yeo, B.L., S. Anastasiadis, S. Kerr and O. Browne (2012), "Synergies Between Nutrient Trading Scheme and the New Zealand Greenhouse Gas (GHG) Emissions Trading Scheme (ETS) in the Lake Rotorua Catchment", Motu Economic and Public Policy Research, Wellington, New Zealand.

Zeitoun, M. (2011), "The Global Web of National Water Security", *Global Policy*, London School of Economics and Political Science, and John Wiley and Sons Ltd.

ANNEXES

Annex A

Glossary of terms*

Acceptability: Risks are acceptable if the likelihood of a given hazard is low and the impact of that hazard is low. There is no pressure to reduce acceptable risks further, unless more cost effective measures become available. See also: *"Tolerability"*.

Disaster: Severe alterations in the normal functioning of a community or a society due to hazardous events interacting with vulnerable social conditions, leading to widespread adverse human, material, economic, or environmental effects that require immediate emergency response to satisfy critical human needs and that may require external support for recovery.

Drought: A period of abnormally dry weather long enough to cause a serious hydrological imbalance.

Exposure: The presence of populations, ecosystems or activities in places that could be adversely affected.

Flood: The overflowing of the normal confines of a stream or other body of water, or the accumulation of water over areas that are not normally submerged. Floods include river (fluvial) floods, flash floods, urban floods, pluvial floods, sewer floods, coastal floods, and glacial lake outburst floods.

Freshwater system: System composed of freshwater and associated aquatic environments in a given geographic area, such as a river basin. The system's hydraulic and biological functions can be modified by human actions on running water, standing water, semi-aquatic and terrestrial elements, both surface and underground, and their interactions. A freshwater system may include one or more ecosystems.

Hazard: An event or a situation with the potential to cause harm.

Impacts: Effects on natural and human systems of contact with hazards.

Likelihood: A probabilistic estimate of the occurrence of a hazard. There are two types of probability: subjective and objective. Subjective or inductive probability is an estimate

---

* This glossary of terms is derived primarily from the International Risk Governance Council's (Renn and Graham, 2006) white paper *Risk Governance: Towards an Integrative Approach* and the glossary of the IPCC (2012) Special Report *Managing the Risks of Extreme Events and Disasters to Advance Climate Change Adaptation* (SREX).

based on the available information and strength of evidence, e.g. taking out insurance. Objective or statistical probability can be calculated where all outcomes are accounted for.[1]

Precautionary principle: A principle employed where the risks of actions or of a failure to act may result in irreversible damage to the environment or other goods. Precautionary measures may be adopted only after a systematic scientific evaluation and must be proportionate, non-discriminatory and duly justified.

Probability: See *"Likelihood"*.

Resilience: Provides the capacity of a (freshwater) system to cope with shocks and undergo change while retaining essentially the same structure and function.[2]

Risk: An uncertain consequence of an event or an activity with respect to something that humans value (definition originally in Kates et al., 1985[3]). Such consequences can be positive or negative, depending on the values that people associate with them. In contrast to "uncertainty", the likelihood of a "risk" can be estimated.

Risk appraisal: The process of bringing together all knowledge elements necessary for risk characterisation, evaluation and management. This includes not just the results of (scientific) risk assessment but also information about risk perceptions (concern assessment) (see Box 1.1).

Risk assessment: The task of identifying and exploring, preferably in quantified terms, the types, intensities and likelihood of the (normally undesired) consequences related to a risk. Risk assessment comprises hazard identification and estimation, exposure and vulnerability assessment and risk estimation (see Box 1.1).

Risk characterisation: The process of determining the evidence-based elements necessary for making judgements on the tolerability or acceptability of a risk. This includes information about economic, social and environmental implications of the risk consequences (see Box 1.1). See also: *"Risk evaluation"*.

Risk distribution: Defines who bears the risk.[4]

Risk estimation: The third component of risk assessment, following hazard identification and estimation, and exposure/vulnerability assessment (see Box 1.1). This can be quantitative (e.g. a probability distribution of adverse effects) or qualitative (e.g. a scenario construction).

Risk evaluation: The process of determining the value-based components of making judgements on the tolerability or acceptability of a risk (see Box 1.1). This includes cost-benefit balancing and incorporation of quality of life implications with a view to maximise expected social welfare. It also includes an assessment of risk-risk trade-offs.

Risk management: The identification of policy options and selection of policy measures to meet the agreed tolerable or acceptable levels of risk in the most cost-efficient manner; the implementation of chosen options and measures and the monitoring of their effectiveness (see Box 1.1).

Risk perception: The evaluation of personal experiences or information about risk by individuals or groups in society.

Risk-risk trade-off: A risk-risk trade-off occurs when interventions to reduce one water risk can increase other water risks.

Risk sharing: The sharing of water risks among stakeholders.[5]

Risk shifting: See *"Risk transfer"*.

Risk transfer: Passing on some or all of the consequences of a risk to a new population, ecosystem or activity, where those benefiting from the risk generating activity are not those who suffer from the risk (e.g. those suffering pollution downstream from a chemical plant).[6] Synonym of *"Risk shifting"*.

Risk trigger: Risk symptoms or warning signs that a risk has occurred or is about to occur.

Target for water risks: Acceptable levels of risk for the four water risks. See also: *"Acceptability"* and *"Water risks"*.

Tolerability: A tolerable risk requires cost-efficient measures to reduce risk to an acceptable level. It may also be tolerable for risk to exceed an acceptable level, provided it is temporary and reversible. See also: *"Acceptability"*.

Uncertainty: An expression of the degree to which a value or relationship is unknown. Uncertainty can result from lack of information or from disagreement about what is known or even knowable.

Vulnerability: The propensity or predisposition to be adversely affected as a result of the exposure to a risk, e.g. structural deficiencies in buildings, vulnerable groups, such as women, children and the elderly.

Water risk: Risk of insufficient water to meet demand in both the short and long-run, including drought; risk of excess water, including flood; risk of water of inadequate quality for a given use; risk of disruption of freshwater systems, when pressure exceeds their coping capacity (resilience). Achieving water security requires maintaining acceptable levels of risk for these four water risks.

Notes

1. See the description of subjective and objective probability provided by UK Climate Projections (UKCP09), *http://ukclimateprojections.defra.gov.uk/21680*.
2. B.H. Walker, N. Abel, J.M. Anderies and P. Ryan (2009), "Resilience, Adaptability and Transformability in the Goulburn-Broken Catchment, Australia", *Ecology and Society* 14(1):12.
3. R.W. Kates, C. Hohenemser and J. Kasperson (1985), *Perilous Progress: Managing the Hazards of Technology*, Westview Press, Boulder.
4. Different meaning as in the insurance industry (the pooling of insurance premiums).
5. "Risk sharing" through capital markets is not covered in this report.
6. Different meaning as in the insurance industry (the shifting of a risk's harmful consequences by way of the insurance contract).

ANNEX B

Basic water facts

This Annex provides basic facts on water supply, demand, quantity and quality at the global and OECD area levels, and projections to 2050. It also provides insights on the potential impacts of climate change and a discussion on indicators to measure water stress.

Water supply

The total quantity of water on earth is around 1.4 billion km^3. However, about 97% of this is stored as salt water in the oceans (Table B.1). Only around 2.5% of all the water on the planet is available as freshwater, of which more than two thirds is stored in ice caps, glaciers, and permanent snow, rendering it more or less unavailable for human use in its present state. Of the remainder, almost all of it is found in aquifers. Of the less than 1% of freshwater that is available in the atmosphere and/or on the surface, two thirds resides in freshwater lakes and less than 1% of this amount is available in rivers which are typically the most easily accessible freshwater source (Anand, 2007).

Table B.1. **Water resources on earth**

		Million km^3	% all water	% freshwater
All forms of water		1 386	100.0	
Seawater		1 351	97.5	
Freshwater	Total	35	2.5	100.0
	Glacial ice, permafrost or permanent snow	24.4	1.8	69.7
	Groundwater and soil moisture	10.7	0.8	30.6
	Freshwater lakes and marshlands	0.1	0.007	0.3
	Rivers	0.002	0.0001	0.01
Evaporation	Off the surface of the oceans	0.505	0.04[2]	
	From land surfaces[1]	0.072	0.21[3]	
Precipitation	On the oceans	0.458	0.03[2]	
	Over land[1]	0.119	0.34[3]	

1. The difference between precipitation onto land surfaces and evaporation from those surfaces is runoff and groundwater recharge – approximately 47 000 km^3 per year (0.13% of freshwater).
2. % of seawater.
3. % of freshwater.
Source: Adapted from Gleick et al. (2009).

Less than 0.2% of the total water on the planet is actually in economic use (Cosgrove and Rijsberman, 2000). This reflects two facts. First, only 2.5% of the water on earth is freshwater. Second, most of freshwater not locked up in ice caps or glaciers comes at the wrong time and place – in monsoons and floods – and 20% is in areas too remote for humans to access.

Only 0.13% of the world's freshwater resources is actually renewable (precipitation onto land surfaces minus evaporation from those surfaces) (Table B.1).

Freshwater is distributed very unevenly. Although 60% of the world's population lives in Asia, the continent has only 36% of the world's water resources (Table B.2). The availability of water per capita varies within the OECD area too (Figure B.1).

Table B.2. **World distribution of freshwater**

Region	% world freshwater	% world population
Total	100	100
North and Central America	15	8
South America	26	6
Europe	8	13
Africa	11	13
Asia	36	60
Australia and Oceania	5	<1

Source: Adapted from WWAP (2003).

Figure B.1. **Renewable freshwater per capita in the OECD area**

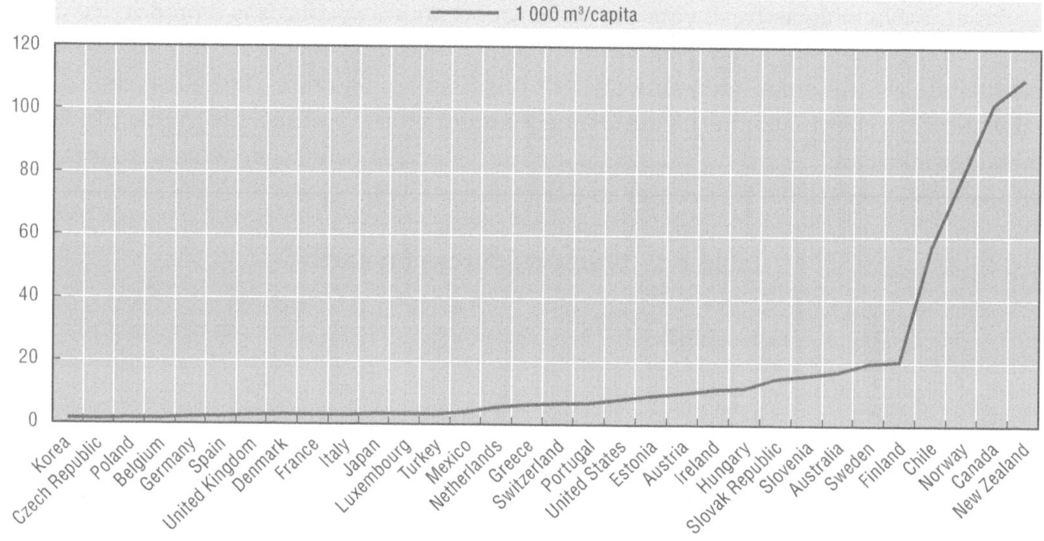

Note: Data refer to year 2011.

1. Total renewable freshwater resources: Net result of precipitation *minus* evapo-transpiration *plus* inflow from neighbouring countries. This definition ignores differences in storage and represents the maximum quantity of water available (long term annual average). When interpreting these data, it should be borne in mind that the definitions and estimation methods employed by member countries may vary considerably among countries.
2. Data for Israel are not available. The data for Iceland is 533 (expressed in 1 000m³/capita).

Source: OECD Environmental Data and OECD Population Data.

Global water resources are governed by the water cycle and the freshwater typically accessible to people consists of two main components: surface water, the water stored on the earth's surface in the form of streams, rivers, and lakes; and groundwater, the water stored underneath the surface in aquifers and underground streams. These two types of freshwater vary in a number of key characteristics and these differences have direct implications for water security.

First, surface water is generally much more accessible than groundwater. Groundwater constitutes nearly 90% of the freshwater on our planet (discounting that in the polar ice caps), but only a small proportion (less than 5%) can be withdrawn each year without depleting the resource base. As a result, less than 20% of total water used globally is from groundwater sources (renewable or not). As a result, although there are much greater quantities of groundwater, over 70% of all fresh water consumed is surface water (Table B.3). Consequently, the availability of surface water is, typically, more important for water security than that of groundwater; although in those regions that experience surface water shortages at least some of these challenges can be mitigated by accessing groundwater. In fact, due to growing levels of surface water scarcity, groundwater is becoming increasingly important, and in certain parts of India and the United States it is the primary source of water used for irrigation.

Table B.3. **Sources of global freshwater use**

	Surface water	Ground water	Drainage water returns	Wastewater reuse	Desalination	Groundwater (non-renewable)[1]	Total
All uses	73	18	5	2	0.3	1	100
Drinking water	48	46	0	0	4	3	100
Irrigation	71	17	7	4	0	1	100
Energy and industry	87	12	0	0	0	0.3	100

1. Nearly 1% of total water used globally (30 km^3 a year) comes from non-renewable (fossil) aquifers mainly in three countries – Algeria, Libya and Saudi Arabia.
Source: Data from AQUASTAT (FAO).

Second, surface water and groundwater differ in the time it takes for the resources to replenish after use. Typically, surface water availability increases much more rapidly and also dissipates more quickly than groundwater. For instance, the global mean time between surface water entering and leaving a system is a little over two weeks, whereas for groundwater reserves it can be up to thousands of years (Oki and Kanae, 2006).[1] As a result, surface water is considered a renewable resource such that the withdrawal of large quantities should have little impact on the amount of water available in the future. By contrast, groundwater may be considered a non-renewable resource depending on the rate of recharge relative to withdrawals.

Third, a further important difference between ground and surface water is the variability of the resource. The lower rate of inflows and outflows to groundwater reserves means that the amount of water is relatively stable even in times of drought, although this depends on the withdrawal rate. By contrast, surface water is driven primarily by variation in precipitation. Therefore, in many locations, especially in semi-arid and arid regions, it does not occur in a regular, predictable way, but varies widely across time and space and may include long periods of drought interspersed with periodic floods (Finlayson et al., 2011). It is this inherent and natural variability in surface water that can cause serious water stress problems at certain times and in particular places.

Thus, the total water resources available in a country are determined by the global water cycle and, in particular, the amount of rainfall each year, as well as inflow from neighbouring countries. While mean or average rainfall helps determine water availability per capita, the variability of that rainfall also plays an important role in determining water stress. Thus, it is important to distinguish between *water resources* and *water supply*.

Total renewable *water resources* are the maximum amount of water that can be abstracted from a natural system, including the total quantity of water in lakes, rivers, aquifers, underground streams, and other sources. However, because not all of this water is accessible for human use, the term *water supply* is typically used to refer to "the amount of water that is accessible to a demand centre and can be delivered reliably and sustainably with respect to the environment or the finite resource base" (2030 WRG, 2009), and it is water supply that is of primary interest when addressing the risks of water shortage, at least in the short and medium term.

The challenge of managing water supply requires balancing the availability of water resources against the demand for water use. For non-OECD countries, this typically involves increasing access to water for basic consumption and sanitation, something that is provided through investment in water supply and delivery. By contrast, in many OECD countries, the vast majority of the population already has access to a reliable water supply, and considerable water infrastructure already exists.

The fact that many OECD residents have ready access to high quality water for drinking and domestic purposes, however, does not mean that water supply is no longer a problem in these countries. Investment in water supply, for instance, is needed to replace ageing infrastructure, meet increasingly stringent water quality controls, and manage extreme weather events (OECD, 2009). Further, growing populations in certain OECD countries, together with migration to locations where water availability is much more variable, such as along the Mediterranean coast or in the Western United States, can exacerbate underlying water supply challenges. As a result of these factors, the number of people affected by droughts in the European Union rose by almost 20% between 1976 and 2006, while the average annual cost of droughts quadrupled to a cumulative cost over this 30 year period of EUR 100 billion (Commission of the European Communities, 2007).

Supply side innovations include desalination water and treated wastewater, which currently account for only 2% of global water use (Table B.3). But desalination is increasingly considered as a viable option for securing water supply. The global desalination market is expected to grow from USD 5-10 billion in recent years to USD 17 billion by 2016 (GWI, 2011). The Middle East and North Africa (MENA) countries currently account for half of the world's desalting capacity for municipal supplies. By 2015, China is expected to become the 2nd largest desalination market in the world after Saudi Arabia. Desalination can rely on a large resource base (oceans) and be widely implemented on coastal areas. The major constraint is that it is energy intensive.[2] Moreover, brine discharges to the sea can increase salinity and temperature, and accumulate toxic compounds, in receiving waters.

There is also a significant potential to increase the share of wastewater being recycled and to widen uses of recycled water – currently used primarily for irrigation – in the industrial and domestic sectors (e.g. for toilet flushing). Industry can be made less dependent on the supply of potable water through obtaining water qualities that are tailored to suit product and process demands and quality standards ("water fit-for-use").

The enormous potential to increase harvested rainwater is largely untapped.

Increasing reliance on groundwater seems problematic without testing the resilience limits of groundwater systems. Moreover, groundwater is particularly vulnerable to long-term, cumulative pollution, which raises substantial uncertainties regarding its future condition. There are prospects to use depleted aquifers to meet strategic long-term water storage of desalinated water.[3]

Water demand

Our knowledge of freshwater use is as poor as our knowledge of freshwater resources – perhaps poorer (WWAP, 2009). Information is largely incomplete, particularly for agriculture, by far the most significant consumer of water, particularly in dry areas where irrigation has been developed. A broad estimate is that less than 10% of the world's renewable freshwater is actually used (Table B.4).

Table B.4. **Freshwater use on earth**

	km^3/year	% renewable water	% water withdrawals
Renewable freshwater[1]	43 569	100	
Total water withdrawals	3 829	9	100
Of which:			
Irrigated agriculture	2 663	6	69
Industry (incl. energy[2])	784	2	21
Domestic (urban)	382	1	10

1. The difference between precipitation onto land surfaces and evaporation from those surfaces.
2. Energy production (hydropower and cooling for thermal stations). Global water withdrawals for energy production were estimated at 583 km^3 in 2010 (IEA, 2012).

Source: Adapted from WWAP (2009).

Irrigated agriculture accounts for nearly 70% of world freshwater withdrawal, while 20% of withdrawals are for industrial uses and 10% for domestic uses (Table B.4).

Water use is measured as *gross* rather than *net* water use. That is, water demand measures the total amount of water withdrawn from the environment (gross demand) rather than the amount of water withdrawn after accounting for the volumes returned (net demand). Gross measures of water use include consumptive and non-consumptive uses of water. By contrast, net water demand only measures consumptive uses of water, so is a measure of *water consumption* rather than *water demand*.

The agricultural sector accounts for more than 90% of annual global freshwater use when considering consumptive uses of water. Because only a part of what is withdrawn is effectively consumed (Table B.5), most of the flow is returned – usually at a lower quality – to the water systems, where it can be reused. Water withdrawals for energy generation – hydropower and thermo-cooling – are on the rise, but energy is one of the economic sectors that consume the least water and it returns most of the water withdrawn back to the water system.

Table B.5. **Global consumption of freshwater withdrawn**

Use	Consumption as % water withdrawals
Domestic (urban)	10-20
Industry	5-10
Energy (cooling)	1-2
Surface irrigation	50-60
Localised irrigation	90

Note: Consumptive use refers to that part of water withdrawn that is evaporated, transpired, incorporated into products or crops, consumed by humans or livestock, or otherwise removed from the immediate water environment.

Source: Adapted from WWAP (2009).

Freshwater use is distributed very unevenly. Withdrawal per capita ranges from 20 m^3 a year in Uganda to more than 5 000 m^3 in Turkmenistan, with a world average of 600 m^3 (WWAP, 2009). In China and the United States, water demands are concentrated in limited parts of the country, in general where agriculture needs to be irrigated or where economic development is occurring.

With rapid population growth and rising incomes, global water withdrawals have tripled over the last 50 years. This trend is explained largely by the rapid increase in irrigation development stimulated by food demand in the 1970s and by the continued growth of agriculture-based economies (WWAP, 2009).

Most OECD countries increased their water abstractions over the 1970s in response to demand by the agricultural and energy sectors. In the 1980s, some countries stabilised their abstractions through more efficient irrigation techniques, the decline of water intensive industries (e.g. mining, steel), increased use of cleaner production technologies and reduced losses in pipe networks. Trends since 1990 indicate a more general stabilisation of water abstractions and a relative decoupling between water use and GDP growth in the OECD area (Figure B.2). About one third of OECD countries have achieved absolute decoupling.

Figure B.2. **Decoupling of freshwater abstraction from GDP in the OECD area**

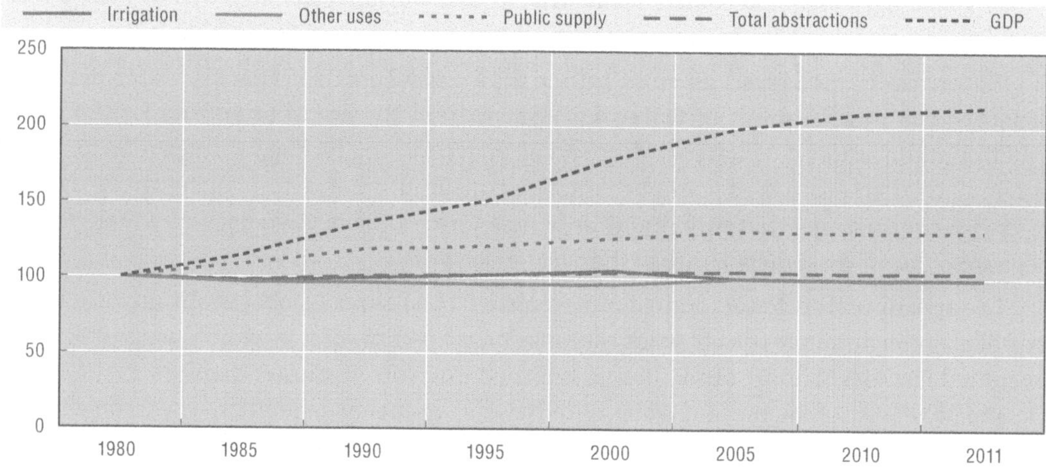

Note: Data exclude Chile, Estonia, Israel and Slovenia.
Source: OECD Environmental Data.

Globally, groundwater withdrawals have risen almost tenfold in the past 50 years, from 100-150 km^3 withdrawals per year to around 950-1 000 km^3 (OECD, 2009). Groundwater withdrawal has increased in many parts of the world, at an alarming scale in India (Shah et al., 2007). Groundwater is a key source of water supply for drinking, irrigation and industrial purposes in many parts of the world. More than 60% is consumed by agriculture in arid and semi-arid regions, producing 40% of the world's food (Morris et al., 2003). Groundwater provides a drinking water source for 60% of the world population (Margat, 2008).

Overall, with the global population projected to grow by 50% between 2000 and 2050,[4] water demand is predicted to grow substantially. This trend, coupled with more variable and possibly reduced freshwater supplies in arid and semi-arid countries, imposes substantial water shortage risks for many nations. Thus, without measures to manage these risks, water scarcity will be a major concern for countries in the OECD and the rest of the world in the future (Anand, 2007; Alcamo et al., 2000; OECD, 2008; OECD, 2012; Raskin et al., 1997; Rijsberman, 2006; Vorosmarty et al., 2000; Vorosmarty et al., 2010).

Water scarcity can be defined as "an imbalance between the supply and demand of freshwater as a result of a high level of demand compared to available supply, under prevailing institutional arrangements (including price) and infrastructural conditions" (Winpenny, 2011). In countries where the limits for extraction of renewable water have largely been reached – and climate change is expected to lower natural water endowments markedly in future – water policies need to switch to demand management, so as to ensure that available resources are put to most efficient and priority use.

Given the widespread under-pricing of water, water demand measures will normally be prioritised and alternative supply options only considered when the potential for water saving and water efficiency increase has been fully exhausted. The choice of whether to increase supply or decrease demand, however, should be decided by cost-benefit analysis.

Water demand is typically divided into three main components: agricultural, industrial, and urban water demand. Of the three components, agriculture makes up by far the largest share of the total, representing about 70% of total global water *demand*.

The sectoral share of water demand, however, is distributed very unevenly. The world can be divided into two groups: in low and middle income countries agriculture is by far the main driver of water demand (80% versus 10% for industry), while in high income countries withdrawals are related mostly to industry and energy (60% versus 30% for agriculture). Urban areas account for a smaller share of water demand (10%) for both groups.

Water will need to feed an additional two billion or so people by 2050; a population increase which will necessitate a large rise in global food production and the water needed to produce it. Food demand is expected to increase by some 70% at the global level by 2050 while it will approximately double for developing countries (FAO, 2011). Global agricultural production is projected to grow at 1.7% annually, on average, for the coming decade (OECD/FAO, 2011).[5]

Increasing prosperity will likely further enlarge the demand for meat which requires more water per unit of production than crops.[6] Indeed food preferences are changing to reflect a world that is rapidly becoming urbanised and wealthier, with declining trends in the consumption of staple carbohydrates, and an increase in demand for luxury products – milk, meat, fruits and vegetables – that are heavily reliant on irrigation in many parts of the world.

However, a decrease in **agricultural water consumption** is expected globally, largely as a result of uptake of improved irrigation technology and, in some areas (e.g. China), lack of land for expanding agriculture. Irrigation currently provides approximately 40% of the world's food, including most of its horticultural output, from an estimated 20% of agricultural land, or about 300 million hectares worldwide. In addition to increasing irrigation water use efficiency, food imports and increases in productivity and production of rainfed agriculture can be mobilised to ensure food security in a context of relative water scarcity (Treyer and Colombier, 2012). Climate change will increase uncertainty on the availability of water for agriculture (e.g. it is expected to alter the seasonal timing of rainfall and snow pack melt and result in a higher incidence and severity of floods and droughts). Both rainfed and irrigated agriculture will need to adapt to reduce resulting production risks (Wreford et al., 2010).

There is substantial scope for water savings in agriculture, where much irrigation water generates little value-added. In many countries, in agriculture, low water prices, combined with the free allocation of water concessions, still hamper an efficient use of water resources. Steps to better take into account water stress in agriculture should include the progressive inclusion of market instruments, such as the tendering of water concessions as well as the elimination of barriers to the exchange of such concessions among users. Water markets are vital for smooth reallocation of water to higher-valued uses elsewhere in the economy and for flexible response to greater hydrological uncertainty. Agricultural water prices will need to rise further so as to reflect service provision costs in full as well as the scarcity and environmental costs of water abstractions (Fuentes, 2011).

Industrial withdrawals are projected to increase as the global economy grows, especially in rapidly industrialising countries such as China and India.

Another significant driver is the growing demand for water in energy production. If today's policies remain in place, in its 2012 world energy outlook the International Energy Agency (IEA) calculates that energy water demand would double by 2035. The largest strain on increased water demand would be soaring coal-fired electricity and the ramping up of biofuel production (Table B.6).[7]

Table B.6. **Energy water demand outlook, 2010-35**

Billion m^3

	2010	2035	2010	2035
	Withdrawal		Consumption	
Primary energy	**38**	**127**	**23**	**64**
Biofuels	25	110	12	49
Coal	2	2	1	2
Gas	2	3	2	3
Unconventional	0.3	1.1	0.3	1.1
Oil	10	12	8	10
Unconventional	0.7	2.1	0.7	2.1
Power generation	**544**	**564**	**43**	**59**
Coal	331	299	38	49
Gas	35	13	2	2
Oil	2	1	0	0
Nuclear	167	222	3	5
Biomass	9	28	1	2
Other renewable	0	1	0	1
Total	**583**	**691**	**66**	**122**

Note: "*Unconventional*" refers to subsets of oil and gas that are considered more difficult or costly to produce. For oil, the subset includes Canadian oil sands, extra-heavy oil, tight oil, gas-to-liquids and coal-to-liquids. In the case of gas, it includes shale gas, tight gas and coalbed methane.
Source: IEA (2012), *New Policies Scenario*.

Steam-driven coal plants always have required large amounts of water. Coal power is increasing in every region of the world except the United States. Coal power producers could cut water demand through use of "dry cooling" systems, but such plants cost three or four times more than wet cooling plants and generate electricity less efficiently.

Biofuels like ethanol and biodiesel account for 18% of energy water demand while providing less than 3% of transport fuels. Future water needs for biofuels depend largely on whether feedstock crops come from irrigated or rain-fed lands and the extent to which second-generation feedstock crops – who tend to be less water-intensive – penetrate markets (OECD, 2010).

Taking all new developments and policies into account, global energy demand is still projected to grow by more than one-third (and energy water demand by about 20%) between 2010 and 2035, with China, India and the Middle East accounting for 60% of the increase in energy demand (IEA, 2012). Energy demand (and energy water demand) will barely rise in OECD countries, where there is a pronounced shift away from oil and coal (and, in some countries, nuclear) towards natural gas and renewable.[8]

Urban water demand is taken to be equivalent to municipal water demand which is defined as the annual quantity of water withdrawn by the public distribution network for direct use by the population (AQUASTAT, 2011). Urban water demand and is also expected to increase as the number of people with access to freshwater supplies grows (particularly in Sub-Saharan Africa) together with increasing levels of urbanisation. In China and India, for example, the proportion of people living in cities is predicted to increase to almost three quarters of their combined populations by 2050, representing a 40% increase from current levels (WEF, 2011).

Cities in certain regions will struggle to find enough water for the needs of their residents and will need significant investment if they are to secure adequate water supplies and safeguard functioning freshwater ecosystems for future generations. In major cities in the developing world, where urban growth is the fastest, modelled results show that currently 150 million people live with perennial water shortage.[9] By 2050, demographic growth is projected to increase this figure to almost 1 billion people (McDonald et al., 2011). Climate change may cause water shortage for an additional 100 million urbanites.

Most studies find that household water demand is fairly price inelastic. However, a survey in 11 OECD countries found that households that face a volumetric charge will, on average, consume about 20% less water than households who do not (OECD, 2011).[10] For those who are charged volumetrically, an increase in the average water price is likely to lead to a reduction in water consumption. The results indicate that a one per cent increase in the average water price across households would lower residential water use by about 0.56%.[11] In Denmark, water price increases have contributed to reduce the average household water use to 110 Litres per head per day (Lhd).[12] At USD 9.18/m^3, the tariff charged to Danish households for water supply and sanitation in Copenhagen is the highest among OECD major cities (IWA, 2010).

Water quantity

Water security is often viewed solely in terms of how much water is accessible to, and is directly used by, people. Water, however, provides for the existence of critically important ecosystem services such as plant growth, natural habitats, nutrient recycling, and waste removal. These processes are essential to the functioning of the natural environment and as a result, water stress can arise when the amount of water in the environment is insufficient for natural environmental processes to function and deliver the services on which people, flora and fauna depend.

A key challenge is to balance water demand for consumptive purposes against the environmental needs for water. Typically, "environmental water requirements" are treated as a residual (King and Brown 2011), and as a result, the lack of water available for environmental needs is creating serious environmental problems. For instance, due to extensive water extractions, some large rivers in semi-arid locations, such as the Colorado and the Murray, only intermittently reach the sea (King and Brown, 2011), and the reduced flows in rivers and water volumes in lakes and wetlands has had a major negative impact on ecosystems. The most infamous example is the Aral Sea, which was once one of the largest freshwater lakes in the world, but is now just 10% of its size as a result of diversions from its main tributaries for irrigation purposes (Mickin, 1988). This transformation has also greatly reduced the water quality of the remaining water making it much more saline. Other forms of degradation have also occurred in many other places including the Murray-Darling Basin (Williams, 2011) and the Mesopotamian wetlands (Partow, 2001).

A recent study by Vorosmarty et al. (2010) found that 65% of all aquatic habitats are under moderate to high threat[13] and this is leading to a freshwater biodiversity crisis; with an estimated 10 000-20 000 freshwater species either threatened or already extinct. The lack of water for environmental needs can also result in wider environmental problems, in the case of the Aral Sea, for example, the reduction in water levels led to salinisation, dust storms, localised climate change, and desertification (Mickin, 1988). Less water in the environment can also impose substantial economic losses on activities dependent on water such as floodplain agriculture and freshwater fishing (both commercial and recreational) (Davis and Hirji, 2003).

Given the expected growth in water demand in many countries, a real concern is that environmental degradation of aquatic habitats will increase. Thus, the management of environmental flows in rivers and lake levels has become an important challenge in terms of water security. The management of flows for environmental purposes includes placing limits on the amount of water which can be extracted from rivers along with restrictions on what water can be used for when extractions occur. The challenge, therefore, is to assess the competing trade-offs between the values of water for consumptive uses versus its value to the environment (Grafton et al., 2011).

Water quality

Water quality is a broad term that refers to how suitable a given volume of water is for a particular purpose or use. Thus, water quality sufficient for drinking purposes would be different to that required for some industrial demands. Typically, water quality is measured by its bio-chemical characteristics that include both organic pollutants and concentrations of chemicals (Day and Dallas, 2011). Quality is affected by readily identifiable sources of pollution such as pulp and paper mills that, typically, discharge pollutants directly into water sources, and also by non-point sources that can only be traced to particular activities, such as farming, rather than specific, identifiable locations.

Each pollutant (organic, chemical) has its own pathway, from its source, through its interactions with the environment, and finally to its effect upon water. Assessing risks of inadequate water quality should thus be site and pollutant-specific.

The three main factors affecting water quality are related to the three principal demands for water: agricultural, industrial, and urban water use. In the agricultural sector, the excessive use of fertilisers to increase yields provides more nutrients than crops can absorb, resulting in excess nutrients leaching into groundwater, lakes and rivers. Untreated slurry from pastoral farming increases the level of pathogens in water sources and can contribute to eutrophication which reduces the amount of available oxygen in the water, with damaging consequences for aquatic flora and fauna (Day and Dallas, 2011). The use of pesticides and herbicides can also release toxic organic chemicals into water sources, some of which can last for decades in the environment (Jones and de Voogt, 1999).

The use of water in industry has multiple impacts on water quality. Large-scale manufacturing and mining can release trace elements and heavy metals such as mercury, zinc, and arsenic into the surrounding water. While such elements can occur naturally in water sources in very small amounts, even slightly elevated levels can be highly toxic. Industrial pollution can also lead to the acidification of water. Mining operations can lead to acid mine drainage whereby sulphate-containing rocks, exposed by the mine, react to form sulphuric acid when in contact with water (Goto, Tanemura and Kawamura, 1978). Likewise, sulphur dioxide, formed by the burning of fossil fuels, can dissolve in water and fall to the earth as acid rain. This can reduce the pH of lakes and rivers with disastrous consequences (Day and Dallas, 2011).

The release of untreated urban sewage into water sources is one of the most important contributors to poor water quality in developing countries; leading to the spread of diseases such as cholera and typhoid (Day and Dallas, 2011). In OECD countries such public health concerns have been addressed with the provision of potable water and the treatment of sewage prior to its release into waterways. However, even with treatment, water quality can be impaired as treated water can contain nutrients which reduce the level of dissolved oxygen in water sources. Another concern is the release of endocrine disrupting compounds (EDCs) such as oestrogen from birth control pills, into water sources. If the subsequent water is recycled for drinking purposes without appropriate dilution, it may impose long-term health risks (Zala and Pen, 2004).

In non-OECD countries, water quality issues are dominated by the need to reduce the incidence of disease associated with consuming contaminated drinking water. By contrast, through large-scale investments in water supply and sanitation, almost all OECD countries no longer face this challenge. Nevertheless, there remain important water quality concerns in terms of both point source (industrial and urban) and non-point source contamination, primarily from agriculture (OECD, 2009).

Projections to 2050

The World Economic Forum's assumption that "as economies grow, more of the freshwater is demanded by energy, industrial and urban systems" (WEF, 2011) is confirmed by the OECD projections, which anticipate that without new policies, allocation of water by 2050 would shift significantly among the main uses (OECD, 2012). The growing demand from manufacturing (+1 billion m^3), electricity (+0.6 billion m^3) and domestic use (+0.3 billion m^3) will compete with water demand for irrigation. As a result, the share of water allocated to irrigation is expected to decline (OECD, 2012).

Overall, the water demand outlook is not optimistic. Under a "business-as-usual" policy context (baseline scenario), a 55% increase in global water demand is projected between 2000 and 2050 (from 3 545 km^3 in 2000 to 5 465 km^3 in 2050) (OECD, 2012). Further, the growth in water demand is not just limited to the use of renewable surface water supplies, and in some countries non-renewable groundwater supplies are expected to be increasingly used at unsustainable rates.

By 2050, the *OECD Environmental Outlook* projects that more than 40% of the world population (3.9 billion people) will live in river basins under severe water stress, meaning in areas where withdrawals exceed safe levels[14] (OECD, 2012). This means an additional 1 billion people compared with today.

Under a "resource efficiency" scenario, the increases in global water demand and water stress would slow down (OECD, 2012). The scenario includes changes in the energy mix (e.g. more solar and wind and less thermal power in electricity generation) and increases in water-use efficiency (e.g. by 15% for irrigation in non-OECD countries; by 30% for domestic uses and the manufacturing sector globally).

The outlook for water quality is not optimistic either. Nutrient pollution from point sources (urban wastewater) and diffuse sources (mainly from agriculture) is projected to worsen in most regions, intensifying eutrophication and damaging aquatic biodiversity (OECD, 2012). Moreover, there is a multiplying number of water contaminants that threaten freshwater quality (including undetected manufactured nanomaterials). For many, discharges are yet to be regulated and wastewater treatment systems are yet to be designed. The expected further degradation of water quality adds to uncertainty about future water availability.

Further shifts in population and greater population densities in low-lying and flood-prone regions heighten the risks associated with flooding. By 2050, flood risks are projected to affect more than 1.6 billion people (nearly 20% of the world's population) and the value of assets at risk will be significantly higher than today (OECD, 2012).

Potential impacts of climate change

In addition to the existing levels of water stress, climate change may impose additional pressures by affecting the demand for water, which is likely to increase with higher temperatures. Global warming is thus likely to aggravate water stress, especially in regions where water is already scarce and where the demand for water is growing rapidly.

While there is much uncertainty over the potential impacts of climate change, there is broad agreement that freshwater resources in certain regions are vulnerable and may be negatively affected (Bates et al., 2008; Lawford, 2011; Vorosmarty et al., 2000). The impacts of climate change on water availability are discussed in the Intergovernmental Panel on Climate Change (IPCC) Fourth Assessment Report. The authors of the report conclude that water availability is likely to increase in wet, tropical areas, but decline in dry, arid areas. The authors also note that Australia, Southern Africa, Central America, the Caribbean, South-western South America, Western United States and the Mediterranean basin are particularly likely to suffer a decrease in water resources. In regions becoming dryer with climate change, the scope for increasing usage of natural water supplies is reduced. The serious shortages of water that are projected in arid regions of the world over the next 50 to 100 years may result in increased frequencies of droughts and water stress (Bates et al., 2008).

The IPCC also predicts that there will be more extreme weather events, leading to increased risks of drought and flooding in some areas. Global warming is accelerating the water cycle (Syed et al., 2010). In a warmer atmosphere, there is likely to be more evaporation and, therefore, more precipitation. More precipitation on land is also likely to lead to more runoff, and thus create more flood risks.

In addition, climate change is likely to affect water quality by inducing physico-chemical, biological and hydro-morphological changes. According to the IPCC Fourth Assessment Report, higher ambient water temperatures, increased precipitation intensity, and longer periods of drought are projected to exacerbate existing water pollution. In addition, if sea levels rise as projected, this will increase the risks of salt-water contamination of freshwater aquifers in coastal areas (Bates et al., 2008).

Measuring water stress

One of the key challenges of water security lies in addressing problems of shortage of water for human and environmental uses. However, due to the significant variations in water availability across space and time, defining terms such as "water stress" are difficult tasks. As a result, there are a number of different ways in which the level of water stress can be measured.

A commonly used **indicator of water stress** is the *Falkenmark indicator* or "water stress index". This approach defines water stress in terms of the relationship between water availability and population; measuring stress as the amount of renewable freshwater available per person per year. According to the water stress index: if the amount of renewable water in a country falls below 1 700 m^3 per person per year, that country is said to be experiencing water stress; if it falls below 1 000 m^3 it is said to be experiencing water scarcity; and absolute scarcity if below 500 m^3 (Falkenmark et al., 1989).

The water stress index is commonly used as a measure of water stress since it is simple, easy to use, and data is readily available. However, it also has limitations in that it: 1) ignores variability in water availability within countries; 2) fails to account for the accessibility of water; 3) does not consider anthropogenic sources of freshwater, such as desalination plants and dams which increase water availability beyond natural flows in a given year; and 4) does not account for the fact that different countries, and regions within countries, have different demands for water (Rijsberman, 2006).

An alternative way of measuring water stress is to use a criticality ratio. This approach relaxes the assumption that demand is uniform across countries and, instead, defines water stress in terms of the relationship between water availability and a country's demand; measuring stress as the proportion of total annual water withdrawals relative to total available water resources (Raskin et al., 1997). Using this approach, a country is said to be "water scarce" if annual withdrawals are between 20-40% of annual supply, and severely water scarce if withdrawals exceed 40%.

The *OECD Environment Compendium* (OECD, 2004) has adopted a similar measure of water stress based on the ratio of water withdrawals (gross abstractions) to annual renewable water (defined as precipitations and inflows from neighbouring countries less evapo-transpiration) (Box B.1). In this approach, if the withdrawal ratio ("intensity of use") is below 10% then water stress is low; if it is between 10-20% water stress is moderate; if the ratio exceeds 20% water stress is medium; and if it exceeds 40%, water stress is treated as severe. While relating supply with demand is helpful, any single index or withdrawal ratio is limited if they: 1) do not consider supply augmentation (such as desalination); 2) ignore withdrawals that are recycled and reused; and 3) fail to consider the capacity of countries to adapt to lower water availability (Rijsberman, 2006).

The causes of water stress differ widely among countries. In Israel, for example, problems of water stress are driven largely by the low levels of water available in the country. According to FAO AQUASTAT data, it has water resources of only 281 m^3 per person per year, a level classified as absolute scarcity using the Falkenmark indicator. Thus, the relatively small natural water supply is the principal cause of water stress. This stress has been mitigated by significant investments in water infrastructure such as desalination plants that produce over 160 million m^3 of water per year, thereby allowing Israel to withdraw more water than is naturally available (AQUASTAT, 2011).

By contrast, in Spain, water stress is more of a problem in terms of overall water demand. With 2 409 m^3 of water available per person per year, Spain would be classified as having no water stress under the *Falkenmark indicator*, however, the demand for water at 728 m^3 per person per year is over four times as large as in Israel and places substantial pressure on the available water resources.

Figure B.1 also highlights the limitations with relying on a national measure of water stress. Despite a rather low intensity of water use at the national level, water stress is a

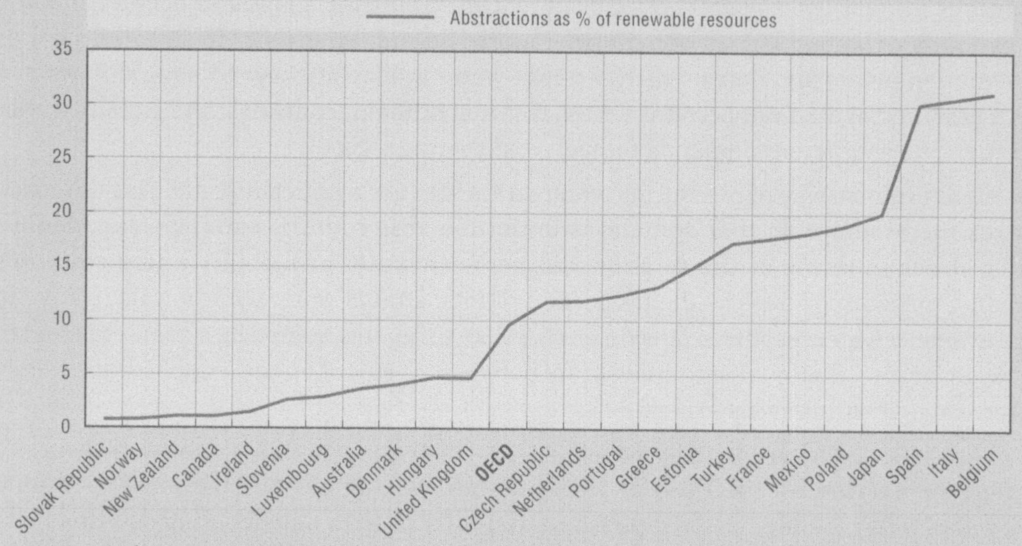

Box B.1. **Water stress in OECD countries**

In 2005, 35% of the OECD's population was living in areas under severe water stress (OECD, 2008). Likewise, Rijsberman (2006) argues that, when taking into account the amount of water required to maintain a healthy environment, large parts of Europe, North America, and Australia face considerable water stress problems. One measure of the extent of water stress in the OECD area is illustrated in the graph below which lists the intensity of water use of OECD countries as measured by the total abstractions relative to total renewable resources.

Note: Data refer to year 2011.

1. Gross water abstraction: water removed from any source, either permanently or temporarily. Mine water and drainage water are included. Water abstractions from ground water resources in any given time period are defined as the difference between the total amount of water withdrawn from aquifers and the total amount charged artificially or injected into aquifers. Water abstractions from precipitation (e.g. rain water collected for use) is included under abstraction from surface water. The amount of water artificially charged or injected are attributed to abstractions from that water resource from which they were originally withdrawn. Water used for hydroelectricity generation is an in-situ use and is excluded. When interpreting these data, it should be borne in mind that the definitions and estimation methods employed by member countries may vary considerably among countries.

2. Data for Austria, Chile, Finland, Germany, Iceland, Israel, Korea, Sweden, Switzerland and the United States are not available.

Source: OECD Environmental Data.

serious problem in OECD countries such as Australia and the United States (the latter not being listed in the figure). Both countries experience serious water stress in certain regions and, as a result, have had at certain times been required to implement water restrictions, establish water markets, and invest in water infrastructure, as well as implementing other initiatives in order to mitigate water stress.

A third measure of water stress has been developed by the International Water Management Institute (IWMI) that includes: 1) water infrastructure, such as water in dams, into the measure of water availability; 2) return flows and recycled water by limiting measurement of water demand to consumptive use rather than withdrawals; and 3) the adaptive capacity of a country by assessing its potential for infrastructure development and efficiency improvements (Seckler et al., 1998). Using this approach the IWMI classifies countries that are predicted to be unable to meet their future water demand without investment in water infrastructure and efficiency as economically water scarce; and countries predicted to be unable to meet their future demand, even with such investment, as physically water scarce (Molden, 2007).

The principal limitation to the IWMI measurement of water stress is that it is data intensive (Rijsberman, 2006). It also fails to consider the ability of a population to mitigate and/or adapt to reduced water availability through the importation of food grown in other countries, or through the use of water saving devices. In turn, this adaptive ability depends on the economic resources available to a country as a whole, as well as to individuals within a country. For instance, the effects of water stress on the poor in developing countries are likely to be much greater than on the wealthy in rich countries.

A fourth approach to measuring water stress, known as the "water poverty index", attempts to capture the income and wealth dimension by accounting for: 1) the level of access to water; 2) water quantity, quality, and variability; 3) water use for domestic, food, and productive purposes; 4) capacity for water management; and 5) environmental aspects (Sullivan et al., 2003). The complexity of this approach, however, means that it is more suited for analysis at a local scale, where data is more readily available, than on a national level. In summary, there is no single measure of water stress which can adequately reflect differences in water supply and demand.

Superficially it may seem that water stress is not a global concern, since less than 10% of the available renewable freshwater resources are withdrawn by humans on an annual basis (Oki and Kanae, 2006); thus, using the OECD indicator of water stress, global water stress would be in the lowest defined category. Further, according to the 2030 Water Resources Group, it appears that most countries have more than enough water to meet the majority of their population's needs without damaging the natural environment (2030 WRG, 2009). However, while these global measures are comforting, aggregate measures which compare total water use relative to the total amount of water available at a global, or even national scale, hide underlying challenges. Such measures are, therefore, of little use for management or policy decision making.

First and foremost, the **problem with global measures** is that water resources are not evenly distributed across the globe and there is substantial variability in water availability across and within countries. For instance, the average annual rainfall in Bangladesh is over 2 600 mm per year, more than 45 times higher than the average rainfall in Saudi Arabia at around 60 mm per year (AQUASTAT, 2011). To illustrate this, Oki and Kanae (2006) show the average annual surface water run-off at a country level. It is apparent that water resources vary considerably between countries, as well as within them. This variability, coupled with the difficulty of transporting water between regions, gives rise to local or regional stress, even when there appears to be adequate total water resources at a global and national level.

Second, water availability within an area or region can vary considerably over time, and even those countries with large water resources can experience water stress at certain times of the year. Bangladesh, for example, has considerable natural water resources due to very high levels of annual rainfall, however, 80% of this rainfall occurs during the monsoon from June to October, leading to floods during the monsoon months and periodic water shortages for the rest of the year (Ahmed, 2011). This variability of water resources over time means that water stress can arise at certain times, whether or not there are sufficient total water resources annually.

Third, in addition to the variability of water resources, demand for water also varies substantially across countries according to various factors such as per capita income, the degree of urbanisation, the structure of the economy, and climatic conditions such as temperature. In 2000, for example, water consumption in Australia was 1 193 m^3 per person. By contrast, water consumption in the same year in France was around 500 m^3 per person (Pacific Institute, 2009). Equally as important, water demand can vary greatly within countries. For instance, water demand in the intensive agricultural zones of the Murray-Darling Basin in Australia is much greater than in the sparsely populated centre of the country. Thus, water stress may arise in areas where local demand exceeds local availability, even if total supplies appear adequate.

Fourth, water stress also depends crucially on the economic resources available to a country, and how much of a country's water resources are accessible for human use. Investment in water infrastructure such as dams, desalination plants, and irrigation, allow countries to manage both the supply and demand for water, thereby allowing greater control of water resources and increasing the amount of freshwater that is accessible. While developed countries have substantial resources available to spend on addressing water issues, developing countries often have far fewer resources, and as a result, water stress can be a much greater problem. In Africa, for example, despite abundant water resources at the continental scale, only 4% of the total available freshwater supply is currently accessed for human use (UN, 2007), thereby creating water stress problems in many African countries.

In closing, worldwide assessments of water stress often rely on fragmented data which is aggregated into country-level statistics and, thus, may fail to identify the extent of water problems (Vorosmarty et al., 2010). Consequently, worldwide assessments of water resources can be misleading and measures of water stress need to be undertaken at a regional scale which accounts for both demand and supply of water, as well as the level of economic development.

Vorosmarty et al. (2000) provide a high resolution analysis of water stress at a regional level by dividing the world into 30' grids in order to capture spatial heterogeneity in water use and availability. Their work maps the distribution of the world's population with respect to relative water demand that is measured as the ratio of water withdrawals from industry, domestic use, and irrigated agriculture (DIA) to the mean annual surface and subsurface (shallow aquifer) runoff (Q) on an annual basis. Their results indicate that in many areas, water stress is a serious concern.

Unsurprisingly, water stress is a serious problem in arid countries with limited water resources, such as Algeria and Saudi Arabia, as well as in densely populated, poor countries with growing water demand and limited resources to exploit available water supplies, such as India and Pakistan. Their study also highlights the fact that water stress is not limited to poor counties or those nations with severely limited water resources, but is also a serious concern in a number of OECD countries such as Australia, Spain and the United States.

Much recent research (Vorosmarty et al., 2000, 2010; Alcamo et al., 2008) has been devoted to illustrating the location and nature of water stress impacts at a global scale. These studies consistently predict that some regions of the world will face water crises: India, northern China, north and sub-Saharan Africa, the Middle East, and parts of Eastern Europe. But these predictions often rely on state variables – climate, precipitation, runoff, population density, aquifer characteristics, land use, and biodiversity – in effect, suggesting that water crises are driven by geospatial factors and therefore are not controllable by human action.

One way to depart from such "top-down" indicators could be to focus more on the human management of water resources, thereby reflecting cultural values, historical context, and political realities. For example, Srinivasan et al. (2012) have grouped 22 basins into six "syndromes": groundwater depletion, ecological destruction, drought-driven conflicts, unmet subsistence needs, resource capture by elite, and water reallocation to nature.

Box B.2. **Asian Development Bank's national water security indicator (NWSI)**

The composite NWSI index is a simple average of 5 sub-indexes. Three sub-indexes reflect water management efforts by household, sectors of the economy, and cities. Two sub-indexes reflect the state of the water environment, expressed in terms of river health as well as resilience to water-related disasters. A score of 1 indicates water insecurity while 5 means the country is a role model for water management.

The household sub-index is a simple average of the scores for three variables: population connected to water supply, population connected to sanitation, population affected by water-related disease (e.g. diarrhea).

The economy sub-index regroups policies in the agricultural, industry and energy sectors (e.g. independence from imported water and goods) and measures the "resilience of economic productivity" (e.g. share of renewable water stored in large dams).

The urban sub-index reflects cities' efforts in terms of water supply and sanitation as well as urban drainage (measured as the economic damage caused by floods and storm water).

The environment sub-index derives from Vorosmarty et al. (2010) and measures the share of basins subject to the following stressors: land use disturbance (e.g. deforestation, road and building, wetland loss); water pollution (e.g. organic, chemical); water resource development (e.g. irrigated cropland, dams); and biotic factors (e.g. non-native fishes).

The last sub-index assesses the country's resilience to three hazards (flood and windstorm, drought, and storm surge and coastal flooding) in terms of exposure, vulnerability and coping capacity (hard and soft).

The recent Asian Development Bank's approach to developing a water security index also focuses on human water management (Box B.2). Overall, the approach is a novel and interesting experiment, but which raises questions and concerns. The implied data collection effort is huge. There are also inadequate caveats surrounding the challenges (e.g. independence from imported goods seems to imply that food self-sufficiency is good for water security, which is not true at all) and limitations associated with making a composite index out of a simple average of 5 sub-indexes covering very different issues. Moreover, it is not clear how an "acceptable" level of water security is defined (although the concept is used).

A move forward would be for countries to identify and map their water risk areas, i.e. the areas where the likelihood and expected impacts of water risks exceed acceptable levels. A national "water security indicator" could then be expressed – for each or all water risks – in physical (e.g. area at water risk as a share of total land area) and financial (e.g. total value of impacts from water risks as a share of GDP) terms. But this requires prior assessment and targeting of water risks, as explained in Chapter 2.

Notes

1. It was estimated that nitrogen released in Lake Rotorua catchment (New Zealand) reaches the lake via groundwater with lags up to 120 years (Rutherford et al., 2011).

2. A recent development in the city of Perth (Australia), where a wind farm provides electricity to run a large-scale desalination plant, could pave the way for the future of desalination.

3. The technology is known as Aquifer Storage and Recovery (ASR).

4. At current fertility levels, the world's population is set to reach 9 billion by 2050 (from today's nearly 7 billion) and could hit 10.1 billion by 2100 (UN DESA, 2011).

5. Compared to 2.6% in the previous decade. Despite the slower expansion, production per capita is projected to rise 0.7% annually.

6. For instance, producing a tonne of beef can require almost 20 times more water than is needed to produce a tonne of maize (Zimmer and Renault, 2003).

7. IEA sees the future impact on water demand of fracking – high-pressure hydraulic fracturing of underground rock formations for natural gas and oil – as relatively small. But fracking may increase risks of water shortage at the local level, and risks of water contamination by methane and fracking fluids.

8. Natural gas power plants use less water than coal plants.

9. Defined as having less than 100 Litres per head per day (Lhd) of sustainable surface and groundwater flow within their urban extent.

10. After controlling for all other potential factors (income, household size, employment, ownership status, residential characteristics, environmental concerns).

11. Water demand of high-income households is less price elastic than the water demand of low and medium-income households.

12. By comparison, England is pursuing the objective to reduce average household water use to 130 Lhd. In Singapore, a national target was set at 140 Lhd by 2030.

13. Vorosmarty et al. (2010) calculate the incident threat level based on multiple stressors that are categorised by four types: catchment disturbance; pollution; water resource development; and biotic factors. Two distinct weighting sets were used to combine the stressors based on expert assessment of stressor impacts on human water security and biodiversity in order to estimate the incident threat from each perspective.

14. River basins with a ratio of withdrawals to available resources that exceeds 0.4.

References

2030 Water Resources Group (2030 WRG) (2009), *Charting Our Water Future: Economic Frameworks to Inform Decision Making*, *www.2030waterresourcesgroup.com/water_full/Charting_Our_Water_Future_Final.pdf*.

Ahmed, K.M. (2011), "Groundwater Contamination in Bangladesh", in R.Q. Grafton and K. Hussey (eds.), *Water Resources Planning and Management*, Cambridge University Press, Cambridge.

Alcamo, J. et al. (2000), "World Water in 2025 – Global Modelling and Scenario Analysis for the World Commission on Water for the 21st Century", *Report A0002*, Centre for Environmental Systems Research, University of Kassel.

Alcamo, J. et al. (2008), "A Grand Challenge for Freshwater Research: Understanding the Global Water System", *Environmental Research Letters*, Vol. 3 No. 1.

Anand, P.B. (2007), *Scarcity, Entitlements and the Economics of Water in Developing Countries*, Edward Elgar, Cheltenham.

AQUASTAT (2011), Main AQUASTAT Country Database, *www.fao.org/nr/water/aquastat/main/index.stm*.

Bates, B.C. et al. (2008), "Climate Change and Water", *technical Paper VI of the intergovernmental Panel on climate change*, IPCC-XXVIII/Doc.13, IPCC Secretariat, Geneva.

Commission of the European Communities (2007), "Communication from the Commission to the European Parliament and the Council Addressing the Challenge of Water Scarcity and Droughts in the European Union", COM(2007) 414 final, CEC, Brussels.

Cosgrove, W.J. and F.R. Rijsberman (2000), *World Water Vision: Making Water Everybody's Business*, World Water Council, Earthscan Publications, London.

Davis, R. and R. Hirji (eds.) (2003), "Water Resources and the Environment", *Technical Note C.1*, Environmental Flows: Concepts and Methods, The World Bank, Washington, DC.

Day, J. and H. Dallas (2011), "Understanding the Basics of Water Quality", in R.Q. Grafton and K. Hussey (eds.), *Water Resources Planning and Management*, Cambridge University Press, Cambridge.

Falkenmark, M. et al. (1989), "Macro-Scale Water Scarcity Requires Micro-Scale Approaches: Aspects of Vulnerability in Semi-Arid Development", *Natural Resources Forum*, Vol. 13, No. 4.

Food and Agriculture Organization of the United Nations (FAO) (2011), "Climate Change, Water and Food Security", FAO *Water Reports*, No. 36, February, FAO, Rome.

Finlayson, B.L. et al. (2011), "Understanding Global Hydrology", in R.Q. Grafton and K. Hussey (eds.), *Water Resources Planning and Management*, Cambridge University Press, Cambridge.

Fuentes, A. (2011), "Policies Towards a Sustainable Use of Water in Spain", *OECD Economics Department Working Papers*, No. 840, OECD Publishing, Paris.

Gleick, P.H. et al. (2009), *The World's Water: 2008-09, The Biennial Report on Freshwater Resources*, Pacific Institute for Studies in Development, Environment and Security, Island Press, Washington, DC.

Global Water Intelligence (GWI) (2011), *Global Water Markets 2011, Meeting the World's Water and Wastewater Needs until 2016*, Media Analytics Limited, Oxford, UK.

Goto, K., T. Tanemura and S. Kawamura (1978), "Effect of Acid Mine Drainage on the pH of Lake Toya, Japan", *Water Research*, Vol. 12, No. 9.

Grafton, R.Q. et al. (2011), "Optimal Dynamic Water Allocation: Irrigation Extractions and Environmental Trade-offs in the Murray River, Australia", *Water Resources Research*, Vol. 47, No. W00G08.

International Energy Agency (IEA) (2012), *World Energy Outlook 2012*, IEA, Paris.

International Water Association (IWA) (2010), *International Statistics for Water Services, Montreal 2010*, Specialist Group Statistics and Economics, IWA, London.

Jones, K.C. and P. De Voogt (1999), "Persistent Organic Pollutants (POPs): State of the Science", *Environmental Pollution*, Vol. 100, No. 1-3.

King, J. and C. Brown (2011), "Inland Water Ecosystems", in R.Q. Grafton and K. Hussey (eds.), *Water Resources Planning and Management*, Cambridge University Press, Cambridge.

Lawford, R. (2011), "Climate Change and the Global Water Cycle", in R.Q. Grafton and K. Hussey (eds.), *Water Resources Planning and Management*, Cambridge University Press, Cambridge.

McDonald, R.I. et al. (2011), "Urban Growth, Climate Change, and Freshwater Availability", *Proceedings of the National Academy of Science of the United States of America* (PNAS), Early Edition. *www.pnas.org/cgi/doi/10.1073/pnas.1011615108*.

Margat, J. (2008), *Les eaux souterraines dans le monde*, BRGM-UNESCO, Paris.

Mickin, P.P. (1988), "Desiccation of the Aral Sea: A Water Management Disaster in the Soviet Union", *Science*, Vol. 241.

Molden, D. (ed.) (2007), *Water for Food, Water for Life: A Comprehensive Assessment of Water Management in Agriculture*, Earthscan/International Water Management Institute, London.

Morris, B.L. et al. (2003), "Groundwater and Its Susceptibility to Degradation: A Global Assessment of the Problem of Options for Management", *Early Warning and Assessment Report Series*, RS. 03-3, UNEP, Nairobi, Kenya.

OECD (2012), *OECD Environmental Outlook to 2050: The Consequences of Inaction*, OECD Publishing, *http://dx.doi.org/10.1787/9789264122246-en*.

OECD (2011), *Greening Household Behaviour: The Role of Public Policy*, OECD Publishing, *http://dx.doi.org/10.1787/9789264096875-en*.

OECD (2010), *Sustainable Management of Water Resources in Agriculture*, OECD Studies on Water, OECD Publishing, *http://dx.doi.org/10.1787/9789264083578-en*.

OECD (2009), *Managing Water for All: An OECD Perspective on Pricing and Financing*, OECD Studies on Water, OECD Publishing, *http://dx.doi.org/10.1787/9789264059498-en*.

OECD (2008), *OECD Environmental Outlook to 2030*, OECD Publishing, *http://dx.doi.org/10.1787/9789264040519-en*.

OECD (2004), *OECD Environmental Data: Compendium 2004*, OECD Publishing, *http://dx.doi.org/10.1787/env_data-2004-en-fr*.

OECD/FAO (2011), *OECD-FAO Agricultural Outlook 2011*, OECD Publishing, *http://dx.doi.org/10.1787/ agr_outlook-2011-en.*

Oki, T. and S. Kanae (2006), "Global Hydrological Cycles and World Water Resources", *Science*, Vol. 313, No. 5790.

Pacific Institute (2009), *Water Data from The World's Water*, *www.worldwater.org/data.html.*

Partow, H. (2001), *The Mesopotamian Marshlands: Demise of an Ecosystem, Early Warning and Assessment Technical Report*, United Nations Environment Programme (UNEP) DEWA TR.01-3, Nairobi.

Raskin, P. et al. (1997), *Water Futures: Assessment of Long-range Patterns and Prospects*, Stockholm Environment Institute, Stockholm.

Rijsberman, F.R. (2006), "Water Scarcity: Fact or Fiction?", *Agricultural Water Management*, Vol. 80.

Rutherford, K., Ch. Palliser and S. Wadhwa (2011), "Prediction of Nitrogen Loads to Lake Rotorua Using the ROTAN Model", *Report prepared for the Bay of Plenty Regional Council*, National Institute of Water and Atmospheric Research, Hamilton.

Seckler, D. et al. (1998), "World Water Demand and Supply, 1990 to 2025: Scenarios and Issues", *International Water Management Institute (IWMI) Research Report*, No. 19, IWMI, Colombo.

Shah, T. et al. (2007), *Groundwater: A Global Assessment of Scale and Significance*, International Water Management Institute (IWMI), 2007.

Srinivasan et al. (2012), "The Nature and Causes of the Global Water Crisis: Syndromes from a Meta-analysis of Coupled Human-water Studies", *Water Resources Research*, Vol. 48, Issue 10, October, W10516, *http://dx.doi.org/10.1029/2011WR011087.*

Sullivan, C.A. et al. (2003), "The Water Poverty Index: Development and Application at the Community Scale", *Natural Resources Forum*, Vol. 27.

Syed et al. (2010), "Satellite-based Global-ocean Mass Balance Estimates of Interannual Variability and Emerging Trends in Continental Freshwater Discharge", *Proceedings of the National Academy of Science of the United States of America (PNAS)*, Vol. 107, No. 42, 19 October, *www.pnas.org/cgi/doi/ 10.1073/pnas.1003292107.*

Treyer, S. and M. Colombier (2012), "Global Irrigation Water Demand Projections to 2050: An Analysis of Convergences and Divergences", April, ENV/EPOC/WPBWE(2012)2.

United Nations (UN) (2007), "Bringing Water to Africa's Poor", *Africa Renewal*, Vol. 21, No. 3.

United Nations Department of Economic and Social Affairs (UN DESA) (2011), *World Population Prospects: The 2010 Revision*, UN DESA Population Division, New York City, *www.unpopulation.org.*

Vorosmarty, C.J. et al. (2010), "Global Threats to Human Water Security and River Biodiversity", *Nature*, Vol. 467.

Vorosmarty, C.J. et al. (2000), "Global Water Resources: Vulnerability from Climate Change and Population Growth", *Science*, Vol. 289, No. 5477.

Williams, J. (2011), "Understanding the Basin and its Dynamics", in D. Connell and R.Q. Grafton (eds.), *Water Reform in the Murray-Darling Basin*, Australia National University E-Press, Canberra.

Winpenny, J. (2011), "Keynote Speech, A Drought in Your Portfolio?", EIRIS Conference, London, 7 June, *www.unesco.org/water/wwap/news/pdf/EIRIS_Conf-JWinpenny_Keynote_Speech.pdf.*

World Economic Forum (WEF) (2011), *Water Security, The Water-Food-Energy-Climate Nexus*, WEF Water Initiative, Island Press, Washington, Covelo, London.

World Water Assessment Programme (WWAP) (2009), *The United Nations World Water Development Report 3: Water in a Changing World*, UNESCO, Paris and Earthscan, London.

World Water Assessment Programme (WWAP) (2003), *The United Nations World Water Development Report 1: Water for People, Water for Life*, UNESCO, Paris and Berghahn Books, New York.

Wreford, A., D. Moran and N. Adger (2010), *Climate Change and Agriculture: Impacts, Adaptation and Mitigation*, OECD Publishing, *http://dx.doi.org/10.1787/9789264086876-en.*

Zala, S.M. and D.J. Pen (2004), "Abnormal Behaviours Induced by Chemical Pollution: A Review of the Evidence and New Challenges", *Animal Behaviour*, Vol. 68, No. 4.

Zimmer, D. and D. Renault (2003), "Virtual Water in Food Production and Global Trade: Review of Methodological Issues and Preliminary Results", in A.Y. Hoekstra (ed.), "Virtual Water Trade: Proceedings of the International Expert Meeting on Virtual Water Trade", *Research Report Series*, No. 12., UNESCO-IHE Institute for Water Education, Delft.

ANNEX C

Costs and distributional impacts of inaction

Drawing on a literature review, this Annex provides examples on the *local* costs and distributional impacts of not managing the risks of water of inadequate quality (microbial and chemical water pollution) and groundwater shortage in selected countries.

Overview

The *OECD Environmental Outlook to 2050* emphasises the consequences of inaction in terms of growing competition among water users to access water resources of adequate quality, growing vulnerability to floods, and increasing pressures on water quality. Inaction (i.e. no new policies beyond those which currently exist) can lead to significant **costs to society and the environment**, with both market and non-market impacts,[1] and can have distributional impacts.

Moreover, water insecurity and commodity production inefficiencies can have *global* impacts (Grey and Garrick, 2012). This is because local water risks may impact on global commodity markets (e.g. a major drought in a food exporting country drives up food prices worldwide) and disrupt supply chains on a global scale (e.g. the 2011 Thai floods led to the closure of multinational electronics and vehicle industries, with impacts cascading through the global economy). The World Economic Forum (WEF) 2011 global risks study identified water insecurity as one of the world's greatest threats with a USD 400 billion annual "risk to business". Water supply crises are among the top 5 risks identified in the WEF Global Risks 2013 survey (by both likelihood and impact). Water insecurity can cause rising material costs, disruptions in the supply chain, increased competition, and regional conflict (Dilley and Hikisch, 2009).

Microbial water pollution remains a major risk in poor countries. The net annual benefit of providing universal access to improved water and sanitation worldwide is estimated at USD 230 billion (see below). In OECD countries, chemical pollution is the main water-related health concern. In Europe, the cost of increased colon cancer from nitrate in drinking water originating from groundwater sources is estimated at EUR 1 billion per year (EUR 4.5 per person) (see below).

The risk of depleting water resources increases where prices of water abstraction do not reflect its scarcity rent and associated environmental externalities. The consequent water shortage directly affects users (domestic, agriculture, industry). It can also have indirect impacts on regional economic activity, such as lost earnings of workers and foregone profits. There can also be important economic externalities, which result in use costs (e.g. subsidence and salination), as well as damages to non-use values, such as the life-support function of water (Figure C.1). In Middle East and North Africa (MENA) countries, the estimated economic cost (foregone extractive uses) of full groundwater depletion is in the range of 1 to 2% of GDP (see below).

Figure C.1. **Social costs of inaction with respect to groundwater management**

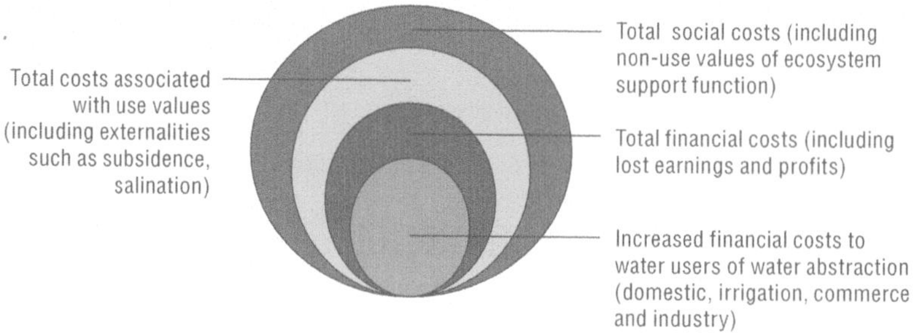

Source: OECD (2008a).

While the economic risks associated with weather-related disasters (e.g. floods, droughts) are only partly attributable to environmental factors, and can only be partly reduced through public policy measures (e.g. flood and drought prevention measures), the costs of inaction in these areas can be considerable. Between 1980 and 2009, global economic losses were estimated at USD 15-30 billion per year for floods and USD 10-15 billion per year for droughts (OECD, 2012). Damage from floods in the United States is now estimated to be USD 5.2 billion each year (Brody et al., 2007).

Water shortages increasingly threaten the viability of energy projects worldwide because the water needed for energy production is set to grow at twice the pace of energy demand through 2035 (IEA, 2012). There is a fear that lack of stable water supplies across the MENA region may lead to future global oil price hikes.

The costs of inaction include direct financial costs associated with environmental degradation (e.g. expenditure on remediation and restoration, private and public health services costs, and private "defensive" expenditure) and other more indirect costs (e.g. the costs of resource depletion and environmental degradation). In addition, costs associated with the loss of environmental use values (which are not reflected in markets) and non-use values (e.g. existence values associated with biodiversity) must be considered.

Freshwater ecosystems depend on a certain level of flows and water quality to function. They already suffer acutely from over-abstraction of water, from pollution of rivers, lakes and groundwater and from poorly-planned water infrastructure. The World Wildlife Fund for Nature (WWF) Living Planet Report shows that declines in freshwater biodiversity are probably the steepest amongst all habitat types. Water shortages also increasingly threaten coastal ecosystems. The coastal zone is one of the most productive ecosystems on earth, and depends vitally on the inflow of clean freshwater (e.g. estuaries, deltas, lagoons, mangrove forests). Maintaining fresh-salt water gradients is a key ecosystem service that produces high biodiversity as well as highly productive fisheries.

Some of the costs described above are already being reflected in household and firm budgets (e.g. expenditure incurred to secure access to clean water) as well as in public sector budgets (e.g. increased public expenditure on health services due to water pollution, dikes and other measures to protect against flooding) (OECD, 2008a). New policies to achieve water security objectives require a careful balancing of the marginal costs of inaction with the marginal costs of further reducing the associated impacts beyond those measures already in place.

There is concern that **segments of the population face greater exposure to water risks** because they are more vulnerable (e.g. children), more exposed (living in areas at risk) and have more limited access to water resources and services (e.g. poorer households).

Globally, the vast majority of premature death from contaminated water, lack of sanitation or inadequate hygiene (about 760 000 people a year in 2011) are children in countries outside the OECD area.[2] Microbial water pollution mostly hurt the rural poor and children (see below). According to the World Health Organization (WHO), 90% of the deaths from unsafe water supply, sanitation and hygiene involve children under five years old. Children are also more vulnerable to water pollution by toxic pesticides, due to high metabolic rates and a reduced ability to detoxify.

Disparities in health risks increase income disparities. Expenditure incurred to secure access to clean water can be a very significant proportion of a household's budget. Because they invest less in health-related water security than higher income groups (e.g. they purchase less bottled water), as it would represent a higher share of their disposable income, lower income groups are more likely to be exposed to water pollution and potentially "pay" a higher share of the health costs of policy inaction.

Groundwater overdraft marginalises farmers who lack capital to invest in well-deepening and those who cannot afford increased water pumping costs (as a result of falling water tables) (see below).

Vulnerability to floods is not evenly distributed within countries and often the poorest suffer disproportionally. For example, Dhaka, Kolkata, Shanghai, Mumbai, Jakarta, Bangkok, and Ho Chi Minh City represent the cities with most people at risk to flooding and all are also situated in countries with low national GDPs per capita (OECD, 2012).

However, assessing distributional impacts should extend beyond assessment of relative levels of exposure to "bads" or access to "goods". It is necessary to look at the underlying demand for environmental quality across different households (Johnstone and Serret, 2006). Not all households have identical preferences and assuming that this is the case can lead to misguided policy conclusions. Depending upon the nature of the income-demand relationship, the distributional effects of inaction may differ if expressed in physical terms only (e.g. levels of exposure) or when underlying preferences (e.g. willingness-to-pay) are also taken into account.

Water of inadequate quality

Microbial water pollution is largely circumscribed to non-OECD countries. Globally, approximately 1.1 billion people lack access to *safe* drinking water[3] and 2.6 billion people to improved sanitation, causing millions of death annually, mainly among young children. The net annual benefit of providing universal access to improved water and sanitation worldwide was estimated at USD 230 billion (Table C.1). This essentially relates to productivity gain and time savings. If only accounting for health benefits, the annual net benefit is USD 0.7 billion.

Improved water supply could decrease world's diarrhoea morbidity by more than 20% (between 21 and 27% according to studies), and improved sanitation by between 22 and 35% (Esrey et al., 1991; Prüss et al., 2002; Fewtrell et al., 2005).

In Mexico, the costs of infectious intestinal diseases caused by untreated water have been estimated at USD 500 million in 2001, including the costs of medical attention as well as productivity loss (Tudela, 2005). In the United States, the provision of safe (treated) drinking water supply in the 20th century has resulted in a net social rate of return on infrastructure investments of 23:1, as well as reduced urban child mortality (Cutler and Miller, 2005).

In Brazil, development of water and sanitation services in the period 1970-2000 has resulted in a net welfare gain of USD 10 300 per capita (Soares, 2007). In India, the health

Table C.1. **Cost benefit analysis of improving water supply and sanitation at the global level per year**

Environmental interventions	Implementation costs (USD billion)	Health benefits (avoided health costs) (USD billion)	BCR (just health benefits)	Total benefits (USD billion)	BCR (total benefits)
Halving the proportion of the population who do not have access to improved water sources and improved sanitation facilities (MDG for water and sanitation)	11.3	14.3	1.3	128.9	11.4
Access for all to improved water and improved sanitation	22.6	23.3	1.0	252.5	11.2
A minimum of water disinfected at the point of use for all, on top of improved water and sanitation services	24.6	77.3	3.1	306.5	12.5
Access for all to a regulated piped water supply and sewage connection into their houses	136.5	100.9	0.7	506.3	3.7

Notes: BCR: Benefit-Cost Ratio. MDG: Millennium Development Goal (UN).
Source: Hutton and Haller (2004), adapted by OECD (2008a).

costs (excluding productivity loss) of water pollution have been estimated at between USD 3 and 8.3 billion annually (Brandon and Homman cited by Maria, 2003).

Unlike microbial contamination, **chemical water pollution** is raising concerns about conformity with mandatory health standards in many OECD countries (Gagnon, 2007). Unlike microbial contamination, the health effects of chemical contamination tend to be chronic. The most frequent pollution cases relate to nitrate, phosphorus and arsenic that can increase risks of, respectively, colon cancer, dental fluorosis and skin infections (Gagnon, 2007).[4]

Some pollutants are present in the natural environment. Arsenic is a case in point; it is present in groundwater in Argentina, Bangladesh, Chile, China, India, Mexico, Nepal and parts of Eastern Europe.[5] But the main sources of chemical water pollution are agriculture, industry and households. For nutrients (nitrogen and phosphorus), diffuse sources of pollution (agriculture and atmospheric deposition) are significant and can be even more significant than point sources (discharges from industry and municipal sewage treatment plants; storm sewer overflow) (Durand et al., 2011) (Figure C.2). Atmospheric deposition can exceed the pollution originating from agriculture. For instance, 60% of the nitrogen load in the North Sea originates from industrial combustion and 40% from agriculture (Hertel et al., 2002).

Figure C.2. **Nitrogen and phosphorus load in selected European catchments**

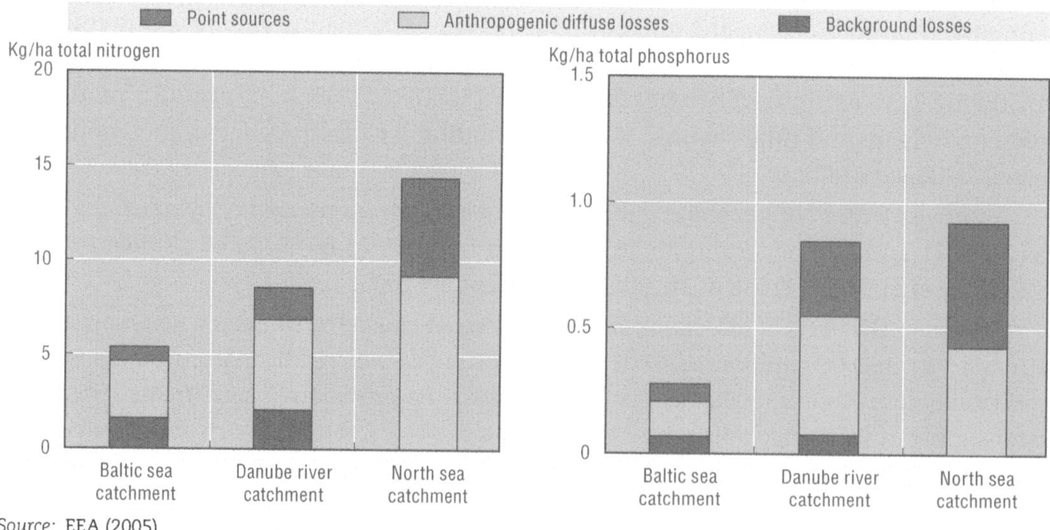

Source: EEA (2005).

Mathaemoglobinaemia[6] is a marginal phenomenon in Europe, where few countries have reported incidence (Albania, Finland, Hungary, Northern Ireland, Romania, Slovakia, Slovenia and Sweden). The highest incidence of mathaemoglobinaemia was reported in Albania and Romania, with 1.25 and 0.75 new cases each year per 100 000 people, respectively (WHO, 2002). In the OECD area, the correlation between the presence of children under 5 in households and the consumption of purified or bottled water is very low, suggesting little concern over mathaemoglobinaemia (OECD, 2011).

On the other hand, there is emerging evidence that the increased incidence of colon cancer is partly due to nitrate in drinking water exceeding 25 mg/L (lower than the 50 mg/L threshold defined by EU's 1998 Drinking Water Directive). In Europe, the cost of increased colon cancer from nitrate in drinking water originating from groundwater was estimated at EUR 1 billion per year (EUR 4.5 per person) (Figure C.3). The cost is likely to be higher when considering drinking water originating from surface waters, as only 5% of the EU population is exposed to high concentration of nitrate (above 25 mg/L) in drinking water abstracted from groundwater, whereas nearly half of the population lives in areas where nitrate exceeds 25 mg/L in surface waters (Grinsven et al., 2006, cited in Grizzetti et al., 2011).

Figure C.3. **Cost of increased colon cancer from nitrate in groundwater-based drinking water in Europe**

EUR million/year

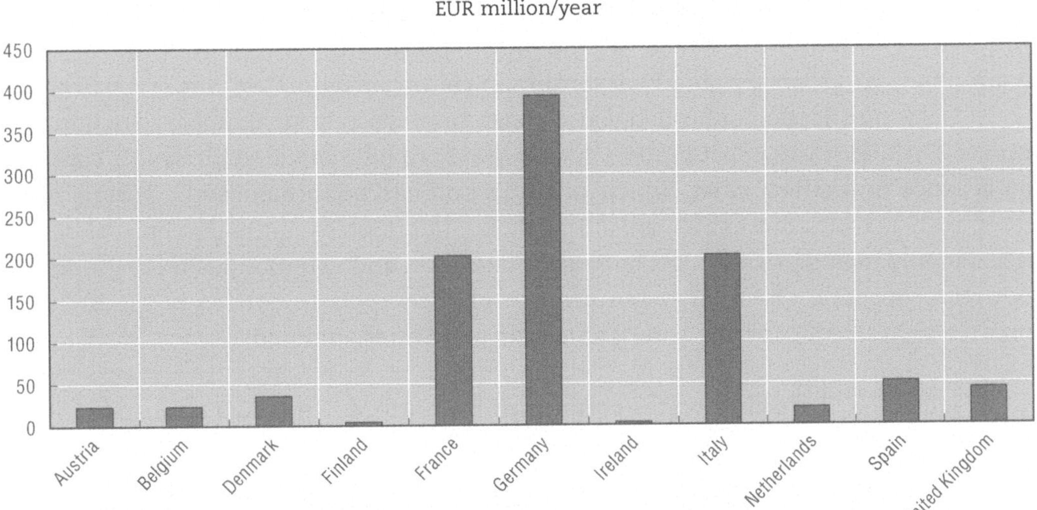

Note: Based on EUR 40 000/yr for premature death and EUR 12 000/yr for health lost.
Source: Adapted from Brink et al. (2011).

Nutrient surpluses can cause water eutrophication, increasing the risk of toxic algal bloom, which in turn can affect human health through the intake of contaminated fish. In the United States, the economic cost of marine harmful algal blooms generated by nutrient runoffs was estimated at USD 49 million, of which USD 22 million of lost wages and work days following ciguatera fish poisoning[7] (Anderson et al., 2000).

Pesticide contamination of water is another important source of health risk. Half of the world population obtains its drinking water from groundwater, where pesticides can remain for decades. In the United States, the environmental and economic costs of groundwater contamination by pesticide were estimated at USD 2 billion (Pimentel, 2005).

In Japan, point-source industrial pollution provoked devastating disease outbreaks in the 1940s and 1950s, such as the Itai-Itai disease due to cadmium and the Minamata disease due to mercury pollution of coastal ecosystems (Kjellström et al., 2006). The net benefit of controlling pollution would have been JP¥ 12 505 (USD 83) million for the Minamata disease and JP¥ 1 910 (USD 12) million for the Itai-Itai disease (Table C.2). The cost of inaction (expressed as the estimated pollution damage costs) includes financial compensations to families (as a proxy to the actual health treatment costs). Most of the health damage costs from the Itai-Itai were borne by farmers of the Jinzu River basin who contracted the disease when irrigated their rice fields with contaminated water (Nogawa and Kido, 1993).

Table C.2. **Cost benefit analysis of past disease outbreaks in Japan**

JP¥ million, 1989 equivalent

Disease	Pollution control cost	Pollution damage costs			
		Health damage	Livelihood damage	Environmental remediation	**Total**
Minamata	125	7 640	4 270	690	**12 630**
Itai-Itai	600	740	880	890	**2 510**

Source: Kjellström et al. (2006).

The cost of policy inaction can also be approached through averting behavior. For example, individuals can invest in private water purifiers or purchase bottled water to avoid the adverse health effects of tap water pollution. In the OECD area, these consumption patterns vary greatly according to country (Figure C.4). But estimating the costs of water purification or bottled water consumption solely attributable to pollution is complex. Purified water consumption is most likely to be motivated by health concerns than reliance on bottled water, which consumption can be influenced by several other factors, such as a lack of trust towards local authorities. Moreover, the purchase of bottled water not only incurs private costs but also social costs and externalities arising from the production process. The energy required to produce plastic bottles is 5.6-10.2 MJ/L compared to 0.005 MJ/L for tap water (Johnstone and Serret, undated).

Figure C.4. **Drinking water consumption patterns in OECD countries**

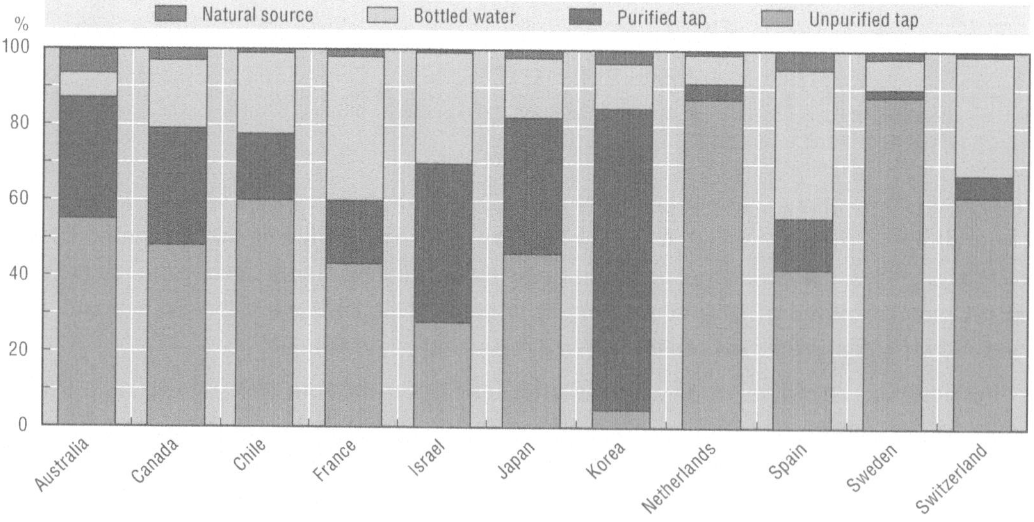

Note: Sample of 12 202 respondents.
Source: OECD (2011).

In France, the cost to households of diffuse pollution from agriculture was estimated at EUR 1-1.5 billion annually (Bommelaer and Devaux, 2011). Of this, EUR 370 million was spent on bottled water or tap water purification and EUR 480-870 million on additional urban water treatment, equivalent to an additional EUR 494 per household per year on the water bill in the most affected regions (Bretagne, Champagnes-Ardennes, Alsace and Garonne).

Beyond health-related impacts, water pollution also affects sectoral activities as well as the environment. Yet, a large part of the cost of policy inaction regarding water pollution has to do with recreational use and ecosystem health losses and the magnitude of these costs might seem insignificant compared with the costs of air pollution where human health impacts dominate (Olmstead, 2010).

Given their strong reliance on natural resources, tourism, fisheries and agriculture tend to suffer most of water pollution. For example, poor sanitation can diminish attractiveness of a country as tourist destination. In Southeast Asia (Cambodia, Indonesia, Philippines and Vietnam), poor sanitation was estimated to cause 5-10% losses in the tourism activity (Hutton et al., 2008).

Agriculture is both a source and a receptor of water pollution. As a receptor, the sector can incur productivity losses due to livestock contamination and reduced plant growth. Toxic contamination of water bodies may poison livestock and generate losses in the meat-processing and milk industries. In Hawaii, 80% of the milk supply was disrupted in 1982 following water contamination by the insecticide heptachlor, generating a loss of USD 8.5 million for the milk industry (Pimentel, 2005). Plants can also absorb dangerous chemicals from contaminated water and pass them on to humans and animals who consume them. Agriculture being a major source of nutrient runoff, rural residents are particularly exposed to health risks from private water supplies.

Commercial and recreational fishing are directly affected by water pollution, as toxic contaminants can kill certain fish species or make them unfit for human consumption. In the United States, harmful algal blooms have an economic cost of USD 18.4 million per year to the commercial fishery sector and USD 6.6 million to tourism/recreation (Anderson et al., 2000). In coastal areas, water pollution can damage coral reef that support sustainable fisheries. In Southeast-Asia, the loss of 1 km^2 of coral reef was estimated to cost between USD 23 000 and USD 270 000, entirely borne by local communities through foregone revenues from tourism, fish export and local fish consumption (Burke et al., 2002).

The manufacturing sector can also be affected. In Australia, eutrophication and algal blooms in the Murray-Darling Basin cost AUD 32.9 (USD 28) million per year, of which AUD 14 (USD 12) million is borne by the manufacturing industry (Atech, 2000).

Eutrophication has a negative impact on waterfront property values (Michael et al., 1996; Krysel et al., 2003; Ara et al., 2006; Dornbusch and Barrager, 1973). Such losses can be estimated through indirect pricing methods, such as hedonic pricing where the value of water quality is capitalised in the value of land. In the United States, depending on the location and size of the property, the losses can amount from a couple of thousands to millions of US dollars per property (Krysel et al., 2003). On the other hand, the benefits of water pollution control on property values in the United States was estimated at USD 1.3 billion for the period 1960-70 (Dornbusch and Barrager, 1973).

There are **disparities at the global level**. For example, surface water quality is correlated with the country average living standards (as expressed by GDP per capita) (Figure C.5). Poverty, the level of country development and water quality depend on each other in a complex

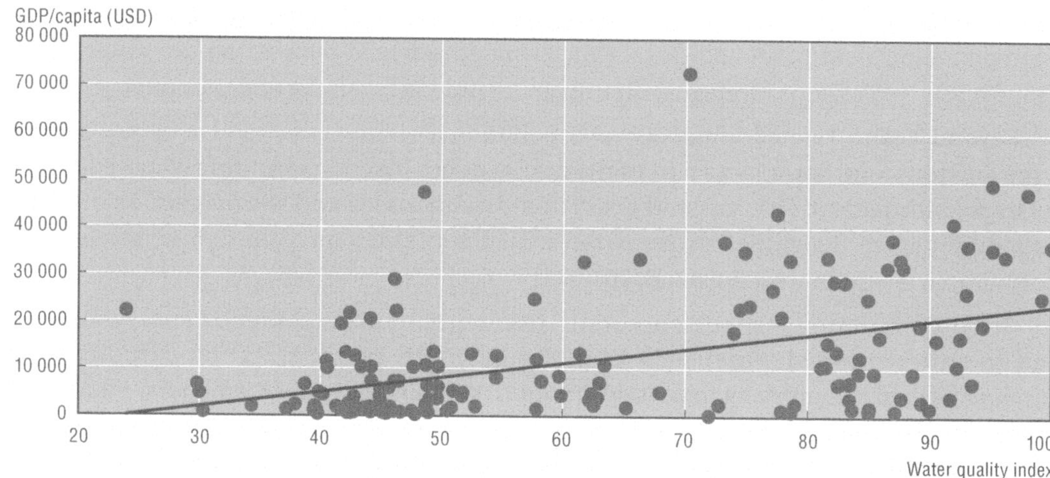

Figure C.5. **Correlation between the level of country development and surface water quality**

Notes: N = 163; Correlation coefficient: 0.4514. GDP/capita is PPP adjusted. Water quality index is a composite indicator taking into account dissolved oxygen, nitrogen, phosphorus, pH levels and conductivity.
Source: Surface water quality is based on GEMS/Water (UNEP) and Yale University, *http://epi.yale.edu/Metrics/ WaterQualityIndex.*

way. For example, the lack of education combined with the lack of water infrastructure increase health risks of polluted water for the poorest segments of the population.

Despite the lack of empirical evidence on socio-economic inequalities in exposure to waterborne diseases, due to the lack of combined epidemiological and socio-economic data, exposure of rural and low-income groups can be assumed to be higher than that of high-income individuals, due to their proportionately greater reliance on common-pool water resources. For example, the poorest farmers in developing countries often abstract water from shallow dug wells (as opposed to deep tube wells) and are thus particularly exposed to water contamination of shallow groundwater. In OECD countries, monitoring of regulated water supplies are less frequent in rural areas than in urban centres and the likelihood that pollution abatement only takes place after a waterborne disease outbreak is thus higher in rural areas.

Health risks from microbial water pollution (e.g. cholera,[8] diarrhoea[9]) increase as levels of income decrease, reflecting the distribution of water supply and sanitation (WATSAN) infrastructure (Figures C.6, C.7 and C.8).

The OECD area accounts for only 1% of the global burden of disease (BoD) attributable to unimproved water supply and sanitation, with Mexico and Turkey having a BoD higher – and a share of their population connected to wastewater treatment lower – than the OECD average (Prüss-Üstün et al., 2004). More than three quarters (76%) of reported deaths attributable to diarrhoeal diseases in the OECD area occurred in Mexico and Turkey.

Microbial water pollution mostly hurt the rural poor. Worldwide, 53% of the rural population lacks access to improved sanitation and 19% to improved water supply. The respective shares for the urban population are 20% and 4% (Figure C.9). Considering that rural areas are in general less well served by health care services and that poverty rates tend to be higher, the related disease-burden is more likely to end in fatal cases.

Figure C.6. **Correlation between cholera, income level and water supply and sanitation in the world**

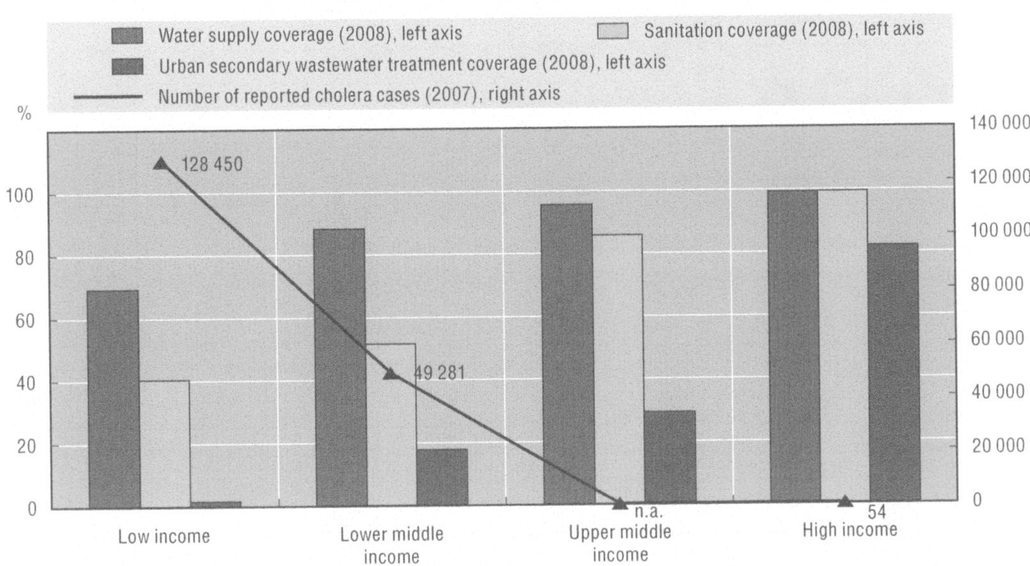

Note: Cholera cases for the "Upper middle income" group are not available in WHO data. The composition of income groups and the affiliation of countries are according to World Bank/WHO criteria.
Source: Data from WHO World Health Statistics 2009; for WATSAN coverage: UN-WHO Joint Monitoring Programmes, Global Water Intelligence and OECD data.

Figure C.7. **Correlation between diarrhoeal disease mortality and country water and sanitation coverage**

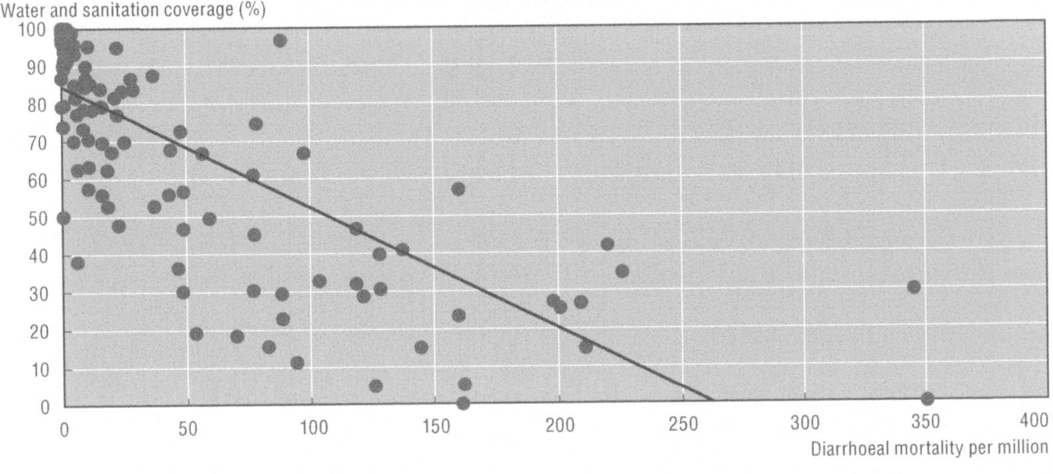

Notes: N = 163; Correlation coefficient: -0.7646.
Source: Water and sanitation coverage is based on Yale University: *http://epi.yale.edu/Countries*. Diarrheal disease is based on WHO Global Burden of Disease data, *www.who.int/healthinfo/global_burden_disease/estimates_country/en/index.html*.

Children (particularly those below the age of 5) are particularly vulnerable to waterborne diseases. According to WHO, children account for most of the death burden of diarrhoeal diseases in the world (Figure C.10).

The OECD area accounts for most of the nitrogen released untreated from sewerage to the environment, reflecting a still low share of tertiary treatment of urban wastewater. But the *OECD Environmental Outlook* to 2050 shows a shifting trend, with non-OECD countries,

Figure C.8. **Correlation between level of country development and water and sanitation coverage**

Water and sanitation coverage (%)

GDP/capita (USD)

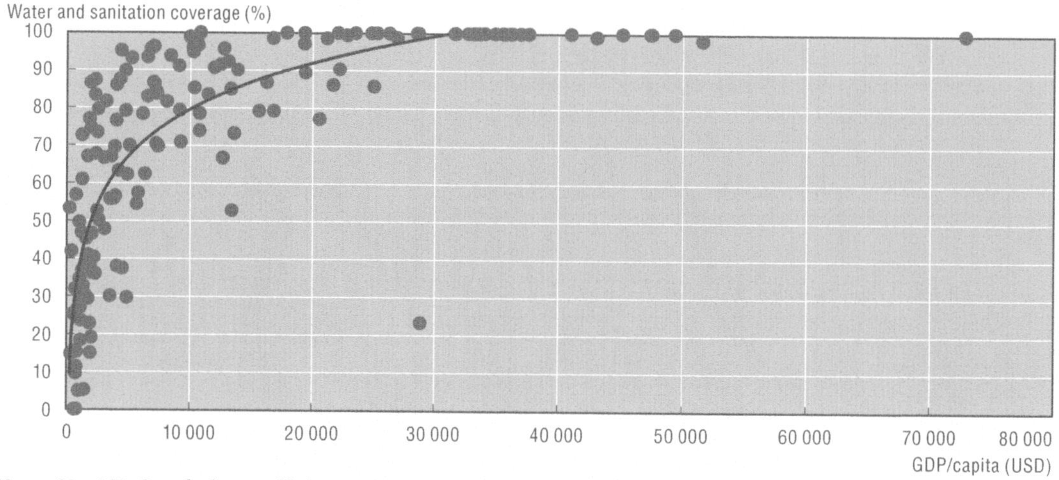

Notes: N = 163; Correlation coefficient: –0.3076. GDP/capita is PPP adjusted.
Source: Water and sanitation coverage index is based on Yale University: *http://epi.yale.edu/Countries*.

Figure C.9. **Improved sanitation coverage in the world**
% of the population

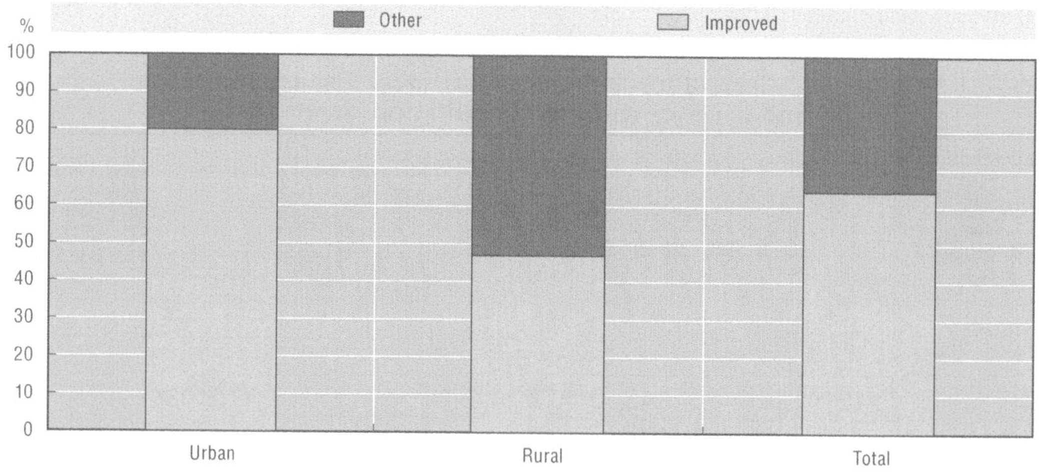

Source: Adapted from WHO/UNICEF (2013).

and particularly China and India, developing urban sewerage faster than (tertiary) wastewater treatment (Figure C.11).

The distributional health impacts of chemical (nutrient) water pollution are likely to be higher in non-OECD economies where, despite growing urbanisation, the rural share of the population is still significant and primarily relies on uncontrolled private wells for drinking water, increasing their exposure to chemical pollution. Children are more vulnerable to water pollution by toxic pesticides, due to high metabolic rates and a reduced ability to detoxify (Pimentel, 2005).

Water pollution can raise inter-generational equity concerns. For example, accumulated groundwater nitrate contamination can last for decades or even centuries even if efficient mitigation measures were implemented now (Durand et al., 2011). The latency period of groundwater pollutants varies according to soil composition and the thickness of aquifers.

Figure C.10. **Estimated deaths from diarrhoea in the world**
Thousand people

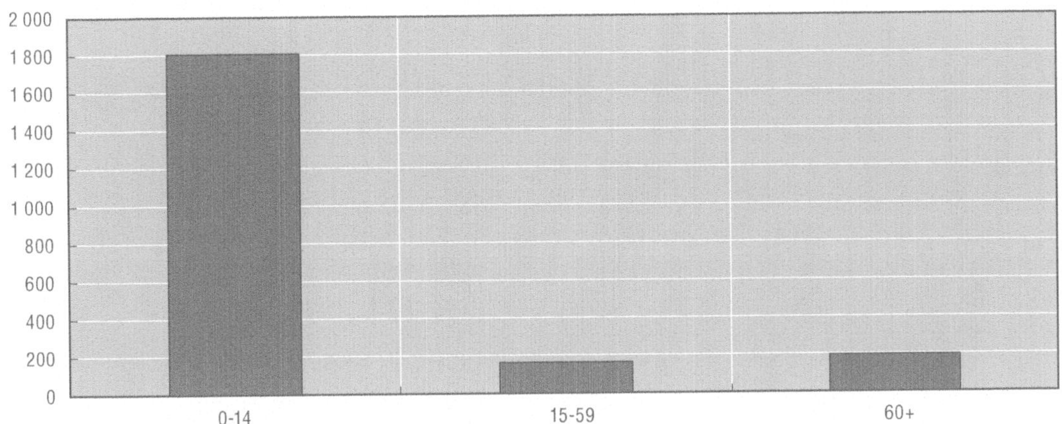

Note: Data refer to year 2004. Deaths from diarrhoea have come down significantly between 2004 and 2011 (see endnote 2).
Source: Adapted from WHO (2008).

Figure C.11. **Nitrogen from untreated urban wastewater in the world**

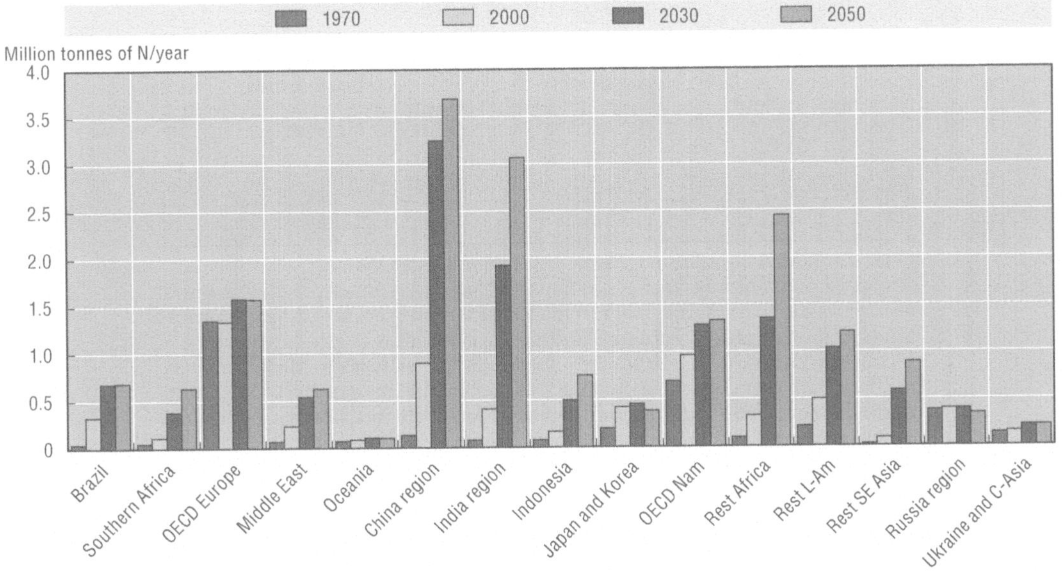

Source: OECD (2012).

There is a similar concern with surface water. In Eastern Europe, despite the halving of nitrogen surpluses following economic transition at the beginning of the 1990s, no significant decrease of nitrogen in surface water has been reported yet (Grizzetti et al., 2011).

There are also **disparities between socio-economic groups at the country level**. For example, 72% of India's population (more than 800 million people) lacks access to improved sanitation. The share is 82% for the rural population against 48% for the urban population (WHO-UNICEF JMP, 2008). There are also inequities in terms of access to safe water, with the rural and the urban poor segments of the population being the most exposed to contaminated water (Figure C.12).

In Southeast Asia, the distribution of unsafe water and sanitation is also unequal between the rural and urban populations, as expressed by the associated economic costs (Table C.3).

Figure C.12. **Access to safe water in India**

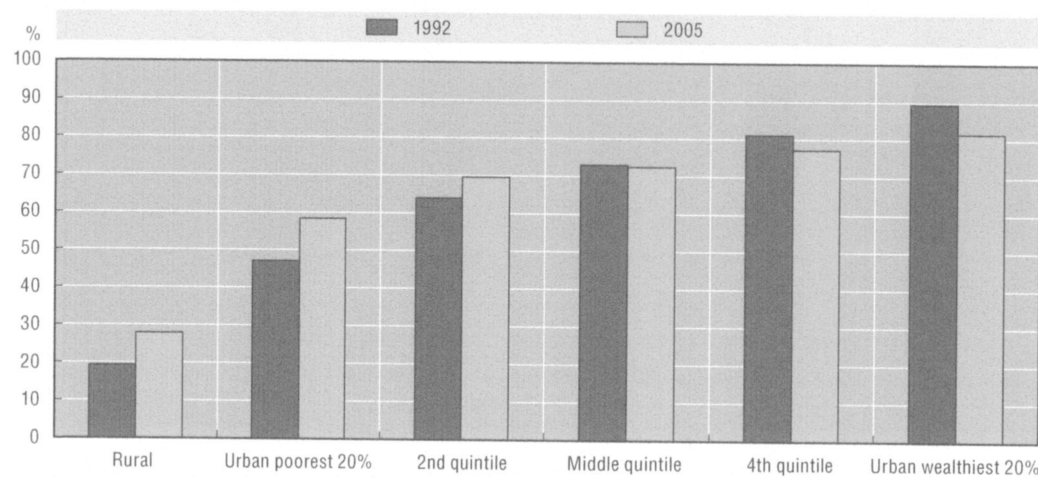

Note: % of households using a piped drinking water source.
Source: WHO, *India: Urban Health Profile, www.who.int/kobe_centre/measuring/urbanheart/india.pdf.*

Table C.3. **Economic costs of unimproved water, sanitation and hygiene in Southeast Asia**

USD million

	Financial costs	Economic costs
Rural	884	4 390
Urban	736	3 532
Non-assigned	407	1 062
Total	2 027	8 984

Notes: Includes Cambodia, Indonesia, the Philippines and Vietnam.
"Financial costs" include changes in household and government spending as well as impacts likely to result in real income losses for households (e.g. health-related time loss with impact on household income) or enterprises (e.g. fishery loss).
"Economic costs" include financial costs plus longer-term financial impacts (e.g. less- and fewer-educated children, loss of working people due to premature death, loss of usable land, tourism losses), as well as non-financial implications (value of loss of life, time use of adults and children, intangible impacts).
Source: Hutton et al. (2008).

In Mexico, about 2% of the national burden of disease (BoD) – or close to 5 000 deaths per year – is attributable to unsafe water and sanitation (and hygiene), affecting mainly children.[10] The poorest states and those where basic sanitation is less developed (Chiapas, Oaxaca, Puebla) bear a higher risk of diarrhoeal disease (Figure C.13).

Child diarrhoeal mortality is also influenced by factors such as the number of doctors/ physicians and health insurance coverage (Reyes et al., 1998). Correlation analysis shows a strong match between child mortality, poverty, the availability of doctors and WATSAN coverage in order of importance (Table C.4). Consequently, rural populations in remote areas are disproportionately at risk.

Disparities in health risks increase income disparities. Several studies – including an OECD report on the health care system of Mexico (OECD, 2005) – show that the poorest segments of the Mexican population have the highest health-related spending in proportion to their revenue (up to 8-9 times higher) (Table C.5).

Figure C.13. **Basic sanitation, poverty and child mortality from diarrhoea in Mexico**

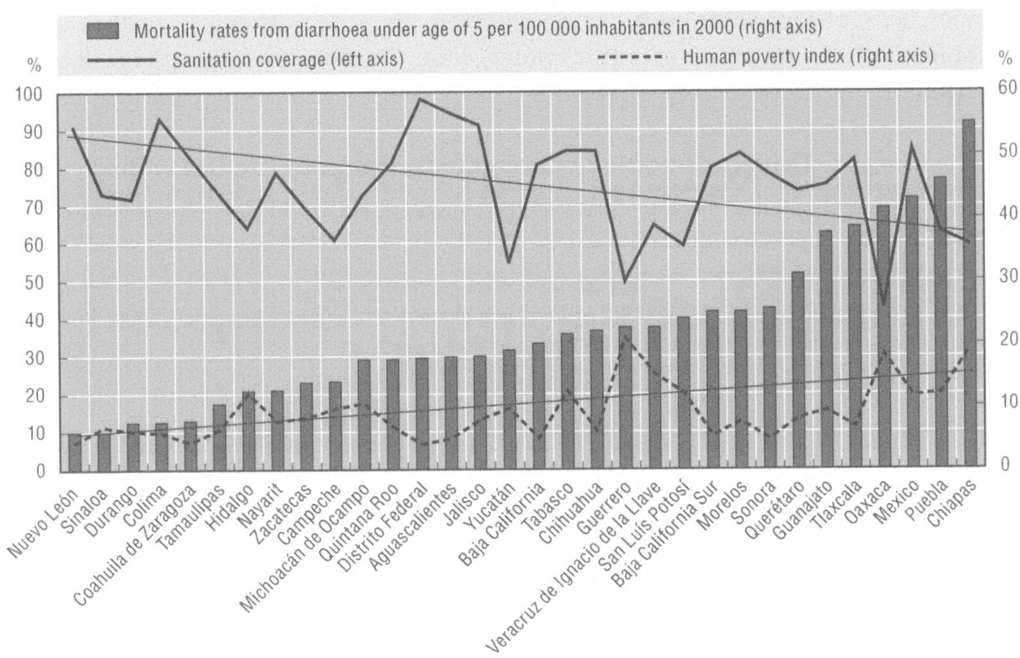

Note: Human Poverty Index is according to United Nations Development Programme (UNDP) definitions and was obtained from CONEVAL. Sanitation coverage was accessed from SEDESOL.
Source: Data compiled from Consejo Nacional de Evaluación de la Política de Desarrollo Social (CONEVAL), Mexico's Ministry of Social Affairs (SEDESOL) and Ministry of Environment and Natural Resources (SEMARNAT). Data refer to year 2000.

Table C.4. **Risk drivers of child mortality from diarrhoea in Mexico**

Risk drivers	Sanitation coverage	Water supply coverage	Human poverty index	Health insurance coverage	Doctor/ 1 000 inhabitants
Correlation with child mortality from diarrhoea (per 100 000 children under 5)	-0.3201	-0.357	0.5386	-0.1082	-0.4403

Source: OECD with data from Mexico's Ministry of Social Affairs (SEDESOL) and Ministry of Environment and Natural Resources (SEMARNAT). Data refer to year 2000.

Table C.5. **Health-related expenditure as a share of income in Mexico**

	By quintile					By insurance status		Total
	I	II	III	IV	V	Insured	Not insured	
Households with catastrophic spending	4.7	4.2	2.4	2.8	2.9	1.2	5.1	**3.4**
Households with impoverishing spending	19.1	0.3	0	0	0	1	5	**3.8**
Catastrophic and impoverishing spending	19.3	4.2	2.4	2.8	2.9	2.2	9.6	**6.3**
Share of households with Social Security	10.8	36.5	50.8	58.3	61.7			

Source: OECD (2005).

In France, 55% of surface waters and 44% of aquifers are reported not to be of good chemical quality (OECD, 2011). A major source of pollution is agriculture. Nitrate in groundwater has been rising over the last decade and in many places it exceeds the 50 mg/L EU threshold (set by the 2006 EU Groundwater Directive), whereas nitrogenous fertiliser and pesticide use are high by European standards.

A recent survey of 11 OECD countries revealed that consumption of bottled water in the OECD area is strongly correlated with high income level, the dissatisfaction over the quality or taste of tap water, and having children under 18. Other key factors are car ownership (eases the transport of bottled water), living in urban area, and the lack of trust in government or local authorities (Table C.6).

Table C.6. **Drivers of bottled water consumption in the OECD area**

	Coefficient	Standard error	Z	P>|z|
Car ownership	0.0484995	0.0186956	2.59	0.009
Living in urban area	0.0820905	0.0332978	2.47	0.014
High income	2.48e-06	6.60e-07	3.75	0.000
Male	0.0037351	0.0310451	0.12	0.904
High education	-0.009305	0.0049857	-1.87	0.062
Age	0.0003677	0.0011889	0.31	0.757
Health concern over tap water	-0.0589095	0.0107405	-5.48	0.000
Taste concern over tap water	-0.1072479	0.0099544	-10.77	0.000
No trust in government	0.0156763	0.0069957	2.24	0.025
Having children under 5	0.0638302	0.0369747	1.73	0.084
Having children under 18	-0.0677674	0.0195465	-3.47	0.001
Individual water metering	-0.0270708	0.0201135	-1.35	0.178
Constant value	0.0710309	0.0952043	0.75	0.456

Note: Correlation degree and logistic regression analysis. The lowest the **P>|z|**, the highest the correlation.
Source: OECD (2011).

Higher income groups tend to purchase more bottled water than lower income groups. Because they invest less in bottled water, as it would represent a higher share of their disposable income, lower income groups are more likely to be exposed to water pollution and potentially "pay" a higher share of the health costs of policy inaction. The same reasoning applies between the urban and the rural population (Figure C.14).

Figure C.14. **Exposure to drinking water pollution in the OECD area**

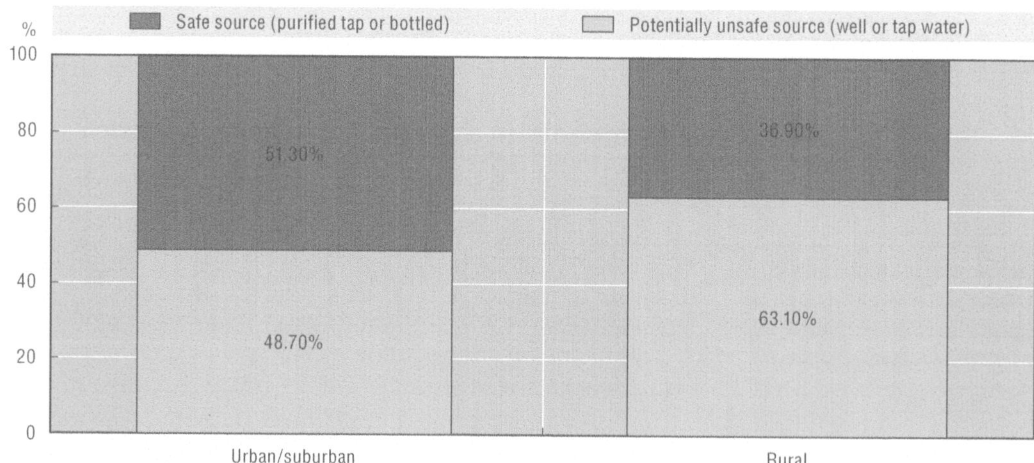

Source: OECD (2011).

Groundwater shortage

Groundwater overdraft defines the condition where the amount of groundwater extracted exceeds the amount of ground water recharging. The economic and environmental costs associated with groundwater overdraft include foregone revenues from extractive uses and the loss of *in situ* values (Table C.7). Most large-scale assessments of the cost of groundwater overdraft are restricted to direct extractive uses. Indeed, the complex relation between aquifers, surface water and wetlands makes any attempt to assess the in-situ values difficult.

Table C.7. **Costing groundwater overdraft**

Groundwater value		Valuation method
Extractive uses	Agricultural use	Derived demand/production cost
	Municipal use	Averting behaviour, contingent valuation
	Industrial use	Derived demand/production cost
In situ values	Ecological values	Production cost techniques, contingent valuation
	Buffer value (resilience to droughts)	Dynamic optimisation, contingent valuation
	Land subsidence avoidance	Production cost, hedonic pricing, contingent valuation
	Recreational value	Travel cost method, contingent valuation
	Existence value	Contingent valuation
	Bequest value	Contingent valuation

Source: Adapted from Freeman (1993) and NRC (1997) (Copyrights 1993 by Resources for the Future).

As regards **extractive uses**, the marginal opportunity cost of groundwater overdraft depends to a large extent on its extractive values and alternative water supplies. In MENA countries, the economic cost (foregone extractive uses) of full groundwater depletion was estimated at 1-2.1% of GDP (excluding *in situ* values) (Ruta, 2005) (Figure C.15). In China, the cost of groundwater depletion (excluding the existence value) was estimated at CNY 92 (USD 12) billion, based on the scarcity value of water (World Bank, 2007).

Figure C.15. **Cost of groundwater depletion as a share of GDP in MENA countries**

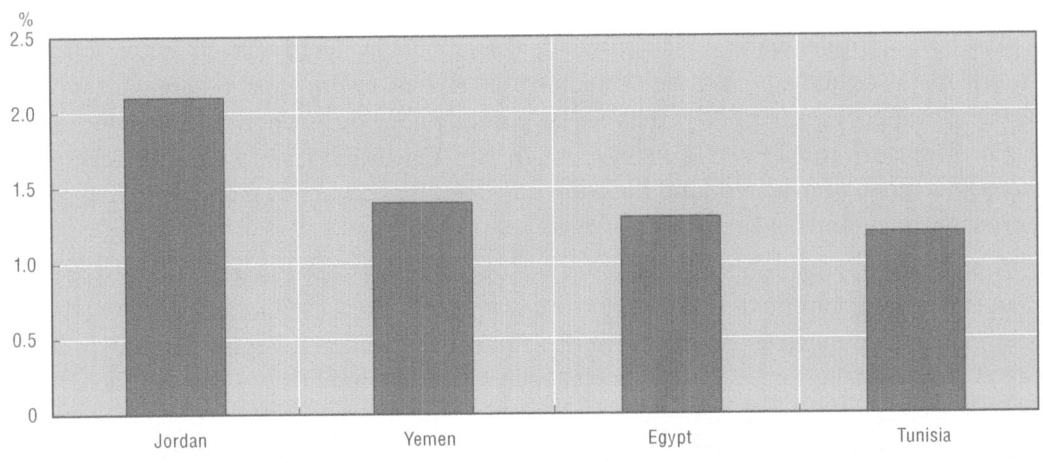

Source: Ruta (2005).

Seawater intrusion caused by groundwater overdraft in coastal aquifers can make groundwater unsuitable for use. This is a major concern in nine of the 11 European countries where coastal groundwater overexploitation was reported (Figure C.16). In the

Figure C.16. **Saltwater intrusion due to groundwater overdraft in Europe**

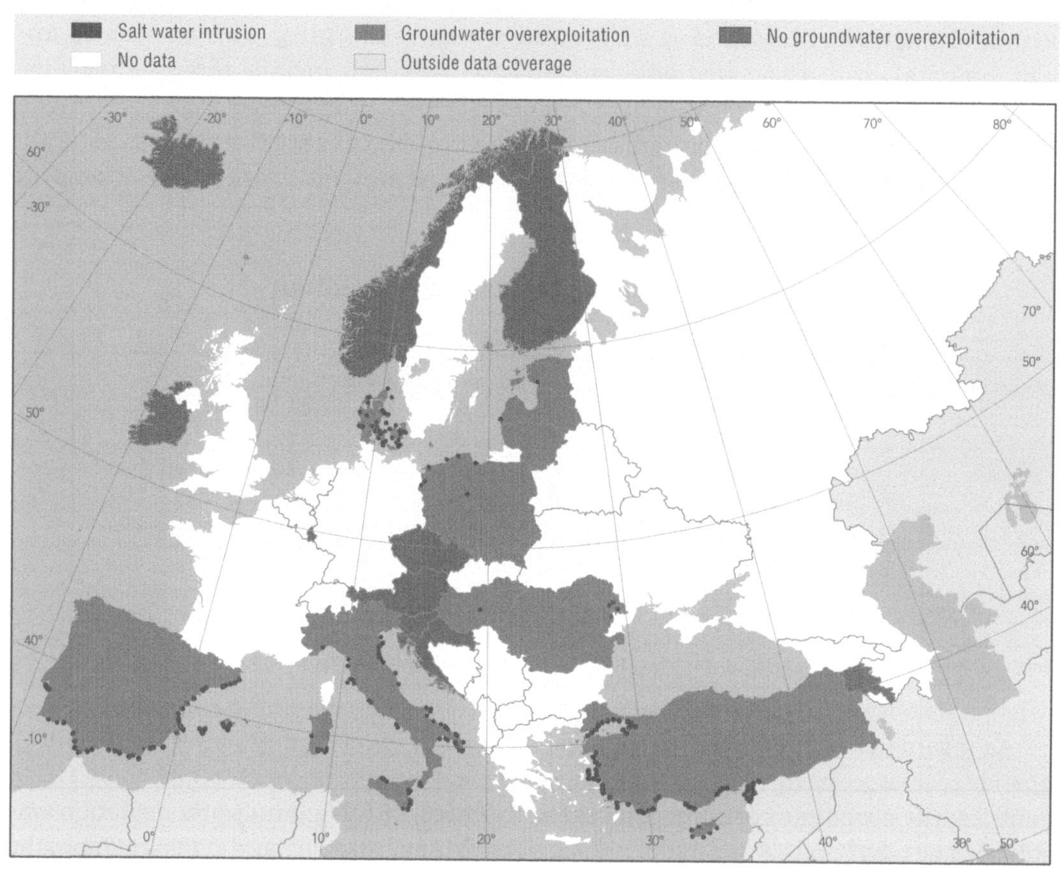

Source: EEA (2003).

Orange County (California), the cost of excessive salinity of drinking water aquifers was estimated at USD 3.41 billion over the 10-year period 1996 to 2006 (NRC, 1997). This accounts for the additional cost of importing water from the Metropolitan Water District.

The cost of unmet water needs is usually the highest in the industrial sector, followed by municipal uses and agriculture (World Bank, 2004). The sector most severely hit is often irrigation, though, given its heavy reliance on groundwater: the economic loss includes lost crop production and lower land values. In the United States, 1 435 million acres (400 000 hectares) of irrigated land have already been lost due to overdraft of the Ogallala aquifer, which extends over portions of eight US states.

Groundwater irrigation contributes up to 10% of India's GDP (Shah, 2007). But several Indian states are confronted with severe groundwater overdraft (Delhi, Haryana, Punjab and Rajasthan) and groundwater development (the ratio of abstraction to replenishment) in Tamil Nadu is at a critical level (85%). This "silent revolution" started in the 1970s and is a direct consequence of unregulated groundwater abstraction, particularly in the irrigation sector. In Karnataka, 20% of the 1.2 million wells go dry each year, representing USD 520 million of lost equipment. Drilling new wells costs INR 8.6 billion (USD 190 million) a year (Shah and Scott, 2004). In Gujarat (India), the foregone revenues from farming as a result of salty irrigation water due to groundwater overdraft was estimated at INR 72 221 (USD 1 550) per acre (Sathyapalan and Iyengar, undated).

Energy policy can significantly contribute to groundwater overdraft, through rebates on diesel tax and electricity tariff when used in irrigation. Indeed groundwater pumping accounts for a significant share of total groundwater abstraction cost. In Mexico, the use of electricity for groundwater pumping in rural areas was reduced along with the reduction of electricity tariff subsidies in the early 1990s (Kemper et al., 2004).

Such energy subsidies not only encourage unsustainable groundwater use but they may also impose a heavy burden on state budgets. In India, electricity subsidies to irrigation largely contributed to the State Electricity Boards' negative return on capital (-39.5% in 2001) (Badiani and Jessoe, 2011). The deadweight loss of electricity subsidies in India was estimated at INR 554 million (USD 13 million) in 1999 (Badiani and Jessoe, 2011). Agriculture accounts for 27-45% of India's energy consumption, depending on states, but only represents 0-12% of total revenues (Shah and Scott, 2004). Electricity subsidies increase the risk of electricity blackouts and other irregularities in service provision, with associated economic costs.

Many urban areas are also heavily dependent on groundwater to satisfy their drinking water needs. Groundwater overdraft and growing urbanisation translate into furthering the distance to find alternative water supplies and associated transfer costs. Mexico City already imports one third of its water from 130 km.

In Texas, groundwater resources are depleting at an alarming rate. By 2060, under business as usual, groundwater availability is expected to decrease by a quarter or 3.6 million acre-feet (4 440 million m^3) according to the Texas State Water Plan 2012 (Figure C.17). The projected economic cost (foregone extractive uses) due to groundwater depletion (mostly in the Ogallala aquifer) is estimated at USD 3 billion per year by 2060 (author's calculations based on data obtained from the Texas Water Development Board). The overall (groundwater as well as surface water) unmet water demand would be 8.3 million acre-feet (10 238 million m^3) by 2060, representing a gross economic cost of USD 115 billion annually.[11]

Figure C.17. **Groundwater overdraft in Texas, 2010-60**

Acre feet

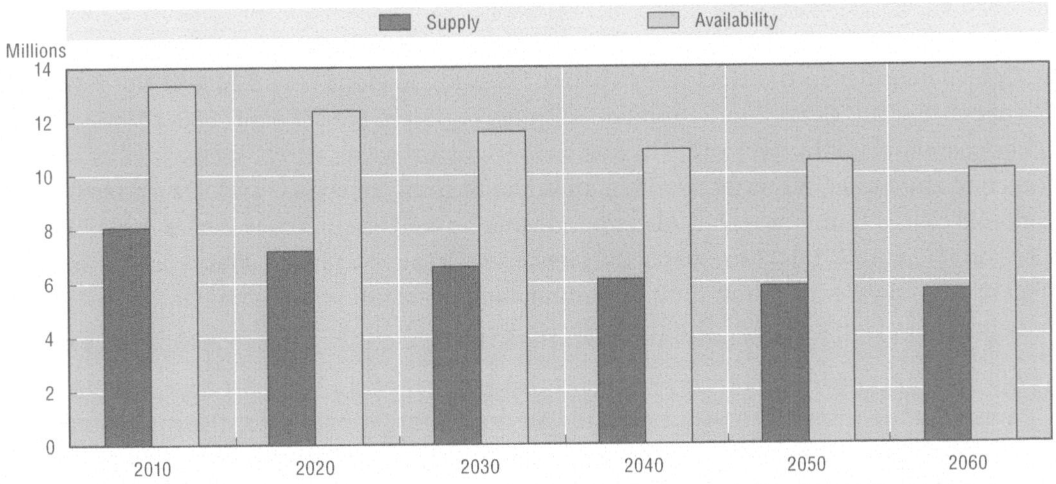

Note: Available grounwater is less than the actual supply due to regulatory restrictions (abstraction rights).
Source: Texas Water Plan 2012 (Draft).

After the devastating droughts of the 1950s, 16 Groundwater Management Areas (GMA) have been established in Texas (Figure C.18). Each GMA must define its "desired future conditions by 2060" based on actual withdrawal data and projections of groundwater availability. A groundwater management plan must be set accordingly.

Figure C.18. **The Groundwater Management Areas of Texas**

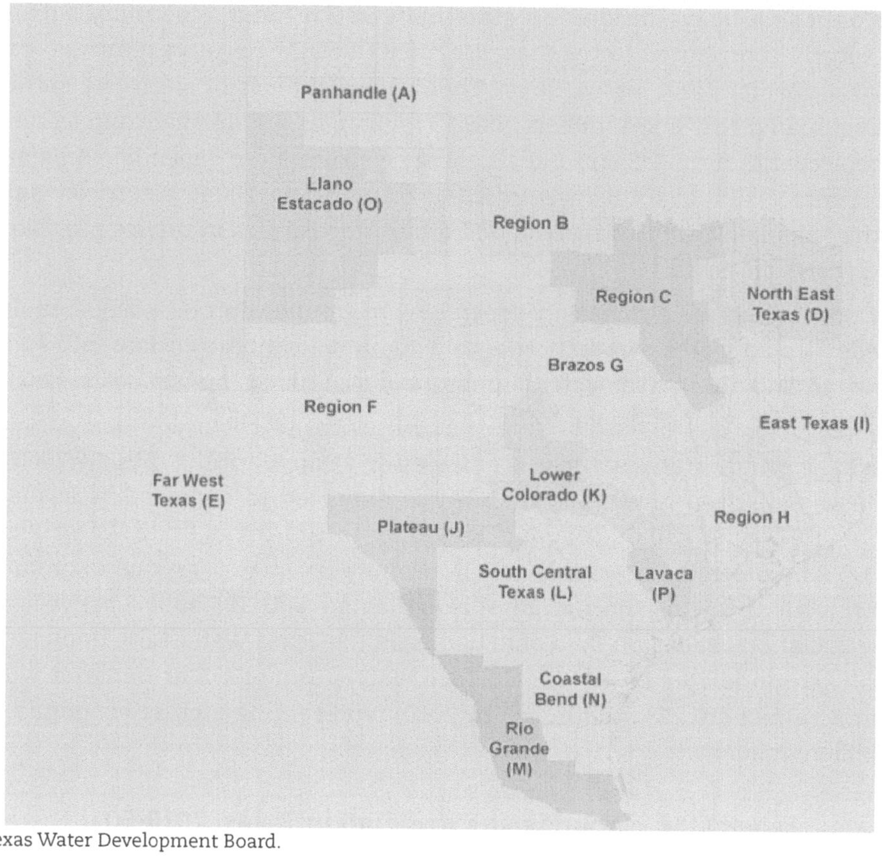

Source: Texas Water Development Board.

The regions of Texas most affected by groundwater overdraft are the High Plains above the Ogallala aquifer in the Northwest (Water Planning Regions O and A) and the Houston area above the Gulf Coast aquifer (Water Planning Region H) (Terrel et al., 2002). Water has a higher economic (extractive) value in Region H (Houston area), where water mostly serves municipal and industrial water needs, than in the farming Regions O and A. As a result, the economic cost of unmet water needs (groundwater as well as surface water) is much higher in Region H (Figure C.19). The regional differences are less pronounced when considering the effect of groundwater overdraft only, as groundwater overdraft is less acute in Region H.

Farming Region A is the most affected when the economic cost is expressed as a share of total regional income (Figure C.20).

Whereas the economic cost of groundwater overdraft is higher for the municipal and manufacturing sectors in absolute terms (Figure C.21), the irrigation sector is the most affected when the economic cost is expressed as a share of total sectoral income (Figure C.22).

Figure C.19. Economic cost of unmet water needs in selected regions of Texas

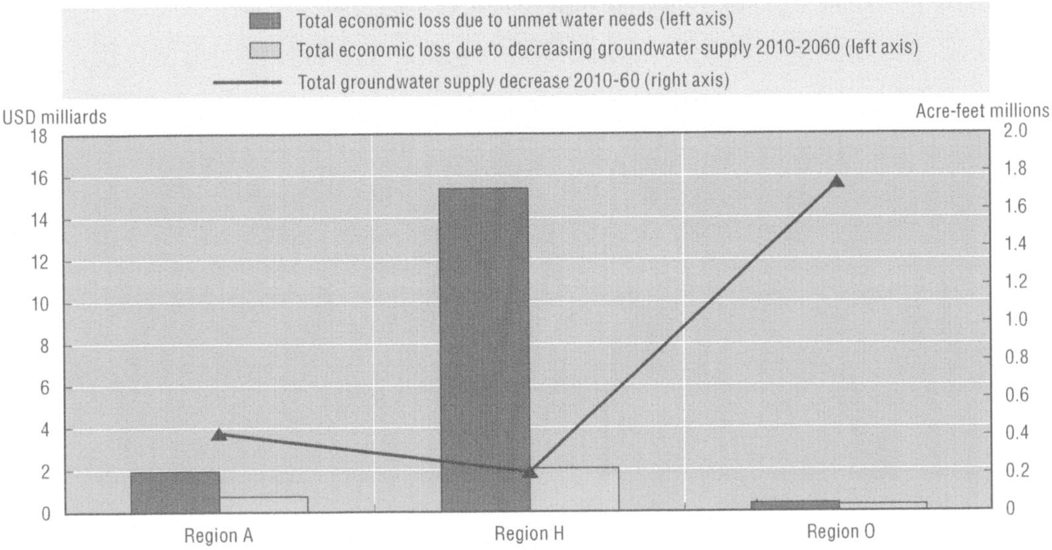

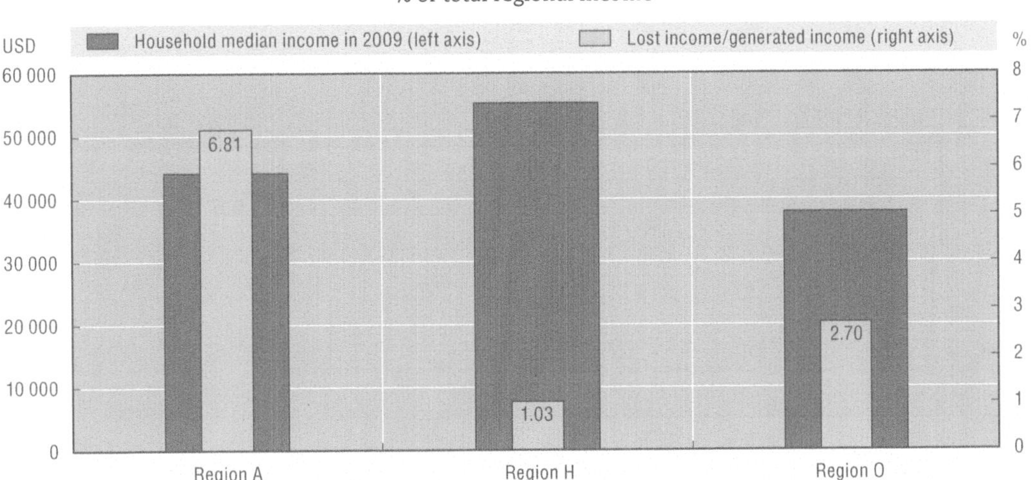

Note: Data refer to year 2060 (outlook under business as usual).
Source: Data from the Texas Water Development Board.

Figure C.20. Economic cost of groundwater overdraft in selected regions of Texas

% of total regional income

Note: Generated income in fiscal year 2000 (latest data available). Data refer to year 2060 (outlook under business as usual).
Source: Data from the Texas Water Development Board.

***In situ* values** also matter. For example, land subsidence caused by groundwater overdraft currently affects many urban areas in the world (e.g. Bangkok, Greater Houston, Mexico City, Osaka, San Jose, Shanghai, Venice). It can negatively impact surface infrastructure (e.g. buildings) and subsurface infrastructure (e.g. cables, pipes, sewerage) and increase flood-prone areas. The cost of inaction will thus differ depending on existing infrastructure, property prices and the area affected.

In the United States, land subsidence affects mainly Texas and California where groundwater overdraft causes more than USD 100 million worth of damage annually (Figure C.23). In the Houston-Baytown area (Texas), the cost of land subsidence caused by groundwater overdraft between 1943 and 1973 was estimated at USD 110 million (including

Figure C.21. **Sector-wise economic cost of groundwater overdraft in selected regions of Texas, USD**

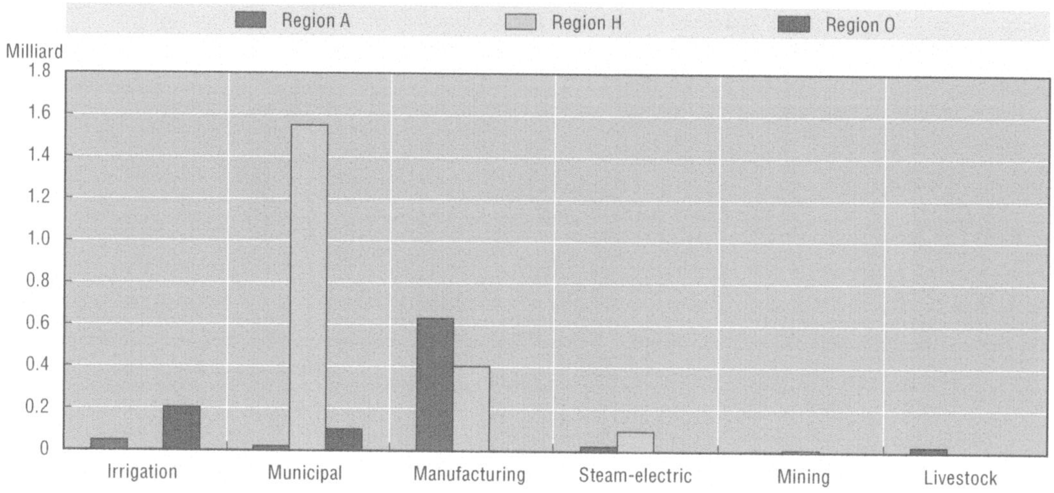

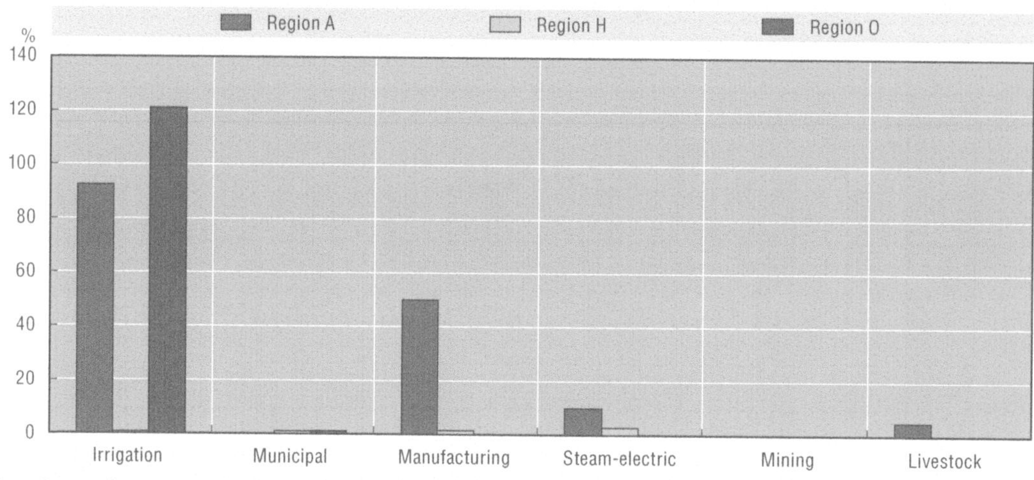

Note: Data refer to year 2060 (outlook under business as usual).
Source: Data from the Texas Water Development Board.

Figure C.22. **Sector-wise economic cost of groundwater overdraft in selected regions of Texas, % of total sectoral income**

Note: Data refer to year 2060 (outlook under business as usual).
Source: Data from the Texas Water Development Board.

damage cost and property value loss) (Warren et al., 1974). Even though this is less than 1% of the total property value, when adding the additional groundwater pumping costs it exceeds the costs of alternative water supply options at the time.

The buffer value of groundwater is the alternative supply that groundwater can provide in case of extended droughts. In the United States it has been estimated at USD 6-8 billion annually (Peters, 2003). In California, the droughts in the early 1990s had a moderate economic impact mainly because farmers could switch from unreliable surface water to groundwater (Gleick and Nash, 1991).

The foregone buffer value due to groundwater overdraft increases with the length of the drought and the related drop in economic output. In Israel, the buffer value of aquifers

Figure C.23. **The cost of land subsidence in the United States**

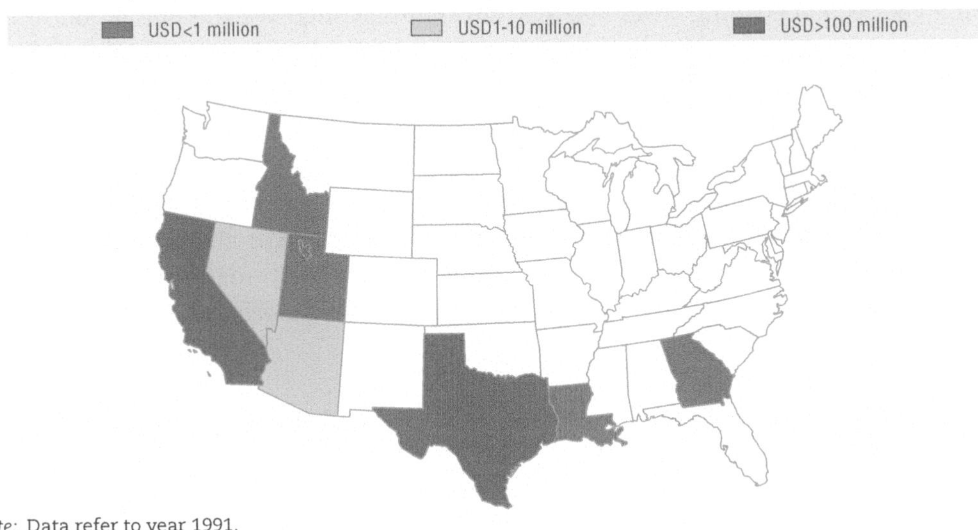

Note: Data refer to year 1991.
Source: Leake (2004).

in the Negev Desert was estimated at more than "twice the benefit due to increase in water supply" (Tsur, 1990).

There are **disparities at the global level**. But distributional impacts of groundwater policy inaction are not necessarily to be expected in major groundwater extracting economies. It foremost depends on the share (and wealth) of the population that rely on groundwater (Figure C.24).

Figure C.24. **Groundwater use in major groundwater abstraction economies**

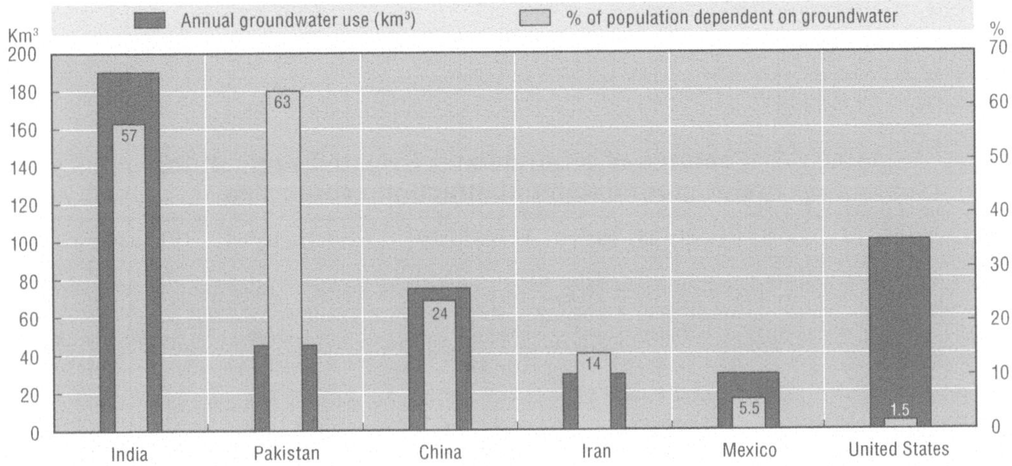

Source: Adapted from Shah et al. (2007).

The opportunity cost of using groundwater now could be inter-generational inequity. When policies fail to internalise the opportunity cost of using groundwater in the future, farmers often have no incentive to take it (i.e. the full cost to society) into account (Ruta, 2005). Under business-as-usual, most of the non-renewable (fossil) aquifers in North Africa and the Sahel region are projected to come to depletion (in terms of exploitable resources) within a timeframe of 50-120 years, which will have a major inter-generational impact. In countries where renewable groundwater aquifers are being overexploited, the higher water pumping and well-deepening costs will also substantially increase the burden on future generations by making groundwater extraction either impossible or unprofitable.

Agriculture is the most important user of groundwater in countries with groundwater overdraft concerns (Figure C.25). The distributional impact of groundwater depletion varies according to the features of groundwater irrigated agriculture (Table C.8). Groundwater-dependent intensive (industrial) farming systems create higher output value per unit of land than family farming systems, but the latter contribute a higher share of GDP and play a major role in poverty reduction. In the family farming systems of South Asia and Northern China, for example, 1.2 billion poor farmers rely primarily on groundwater for their daily income (Shah, 2007).

Figure C.25. **Groundwater withdrawal by sector in major groundwater abstraction economies**

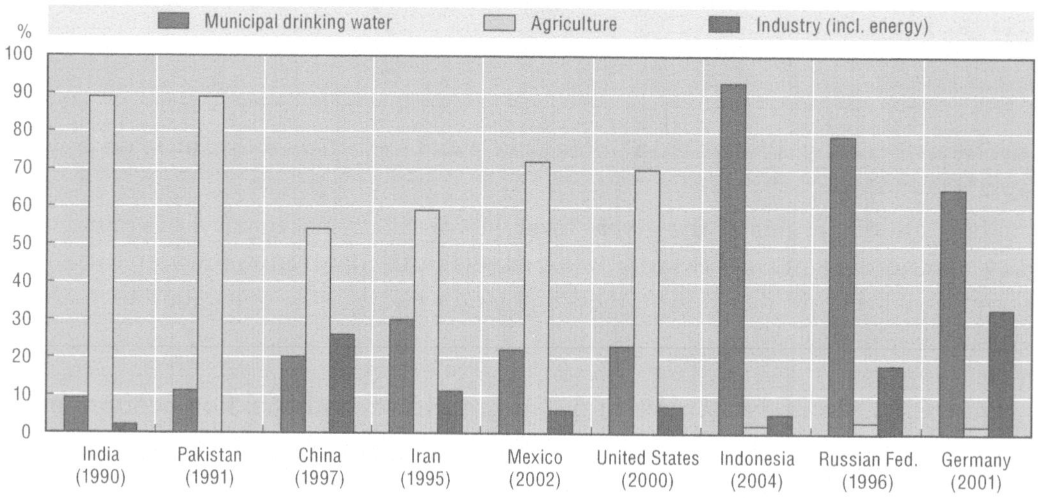

Note: There is no overdraft concern in Indonesia, Russia and Germany.
Source: Adapted from Margat (2008).

Table C.8. **Features of groundwater irrigated agriculture in major groundwater abstraction economies**

Farming system	Farming in arid areas	Intensive farming	Family farming	Extensive pastoralism
Countries	Algeria, Egypt, Iraq, Iran, Libya, Morocco, Tunisia, Turkey	Australia, Brazil, Cuba, Italy, Mexico, Spain, United States	Afghanistan, Bangladesh, North China, India, Nepal, Pakistan	Botswana, Burkina Faso, Chad, South Africa, Tanzania, Zambia
Contribution to GDP	2-3%	Less than 0.5%	5-20%	5-20%
Contribution to national welfare[1]	Low to moderate	Low to very low	Very high	Low
Contribution to poverty reduction	Moderate	Very low	Very high	Low but essential
Gross output value (USD billion)	6-8	100-120	100-110	2-3

1. As expressed by the share of rural population and of food production relying on groundwater.
Source: Adapted from Shah et al. (2007).

There are also **disparities between socio-economic groups at the country level**. For example, when confronted with surface water shortage, farmers tend to shift to groundwater. In that sense, groundwater irrigation can be said to promote equity in access to water. On the other hand, groundwater overdraft marginalises farmers who lack capital to invest in well-deepening and those who cannot afford increased water pumping costs (as a result of falling water tables). Small farms are often more significantly affected by groundwater depletion than large farms. In the state of Punjab (India), the decrease in farm return as the groundwater table falls is more pronounced for small farmers (Figure C.26).

Figure C.26. Return to cost ratios of rice and wheat cultivation in Punjab, India

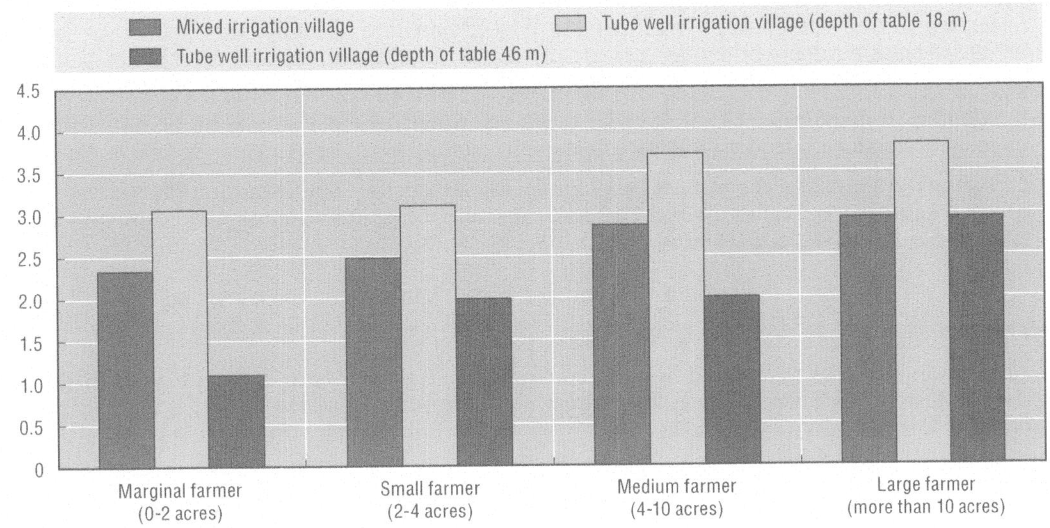

Source: Adapted from Sarkar (2011).

There are similar findings in other Indian states (Nagaraj et al., 2003; Reddy, 2003; Nayak, 2009). In the state of Karnataka, for example, the unit cost of coping with groundwater overdraft is higher for small farmers (Table C.9).

Table C.9. Cost of coping with groundwater overdraft in Karnataka

2001 INR

Coping mechanism	Small farms	Medium-size farms	Large farms
Deepening of well	8 000	10 511	11 808
Additional well	42 500	33 000	40 000
Conveyance pipes from distant borewell	3 536	4 565	15 208
Drip irrigation (per acre of coconut garden)	10 000	10 000	11 875
Coping cost per acre of Gross Irrigated Area	18 762	12 109	14 320

Source: Nagaraj et al., 2003.

Notes

1. The costs associated with the continued loss of biodiversity are likely to be very significant, but their impacts (in terms of lost welfare) are not reflected in market prices.

2. UNICEF Press Release, 22 March 2013. *www.unicef.org/media/media_68359.htm.*

3. Note that the number of people lacking access to safe water is higher than the number lacking access to basic water supply.

4. Other, more episodic, pollution cases involve aluminium (causing neurodegenerative diseases) and lead (endemic diseases of the nervous system).

5. Fluoride is another naturally occurring contaminant in China, India, Sri Lanka and parts of Africa (Kjellström et al., 2006).

6. The health impacts from nitrate pollution are *mathaemoglobinaemia* (blue baby syndrome) and colon cancer. There is a growing scientific consensus, however, that association with the former is relatively weak.

7. Ciguatera fish poisoning (or ciguatera) is an illness caused by eating fish that contain toxins produced by a marine microalgae called *Gambierdiscus toxicus*.

8. Cholera is the only bacterial disease exclusively attributable to water pollution.

9. According to the World Health Organization (WHO), diarrhoea accounts for most of the water-related disease burden worldwide. It is mainly caused by cholera, typhoid fever, shigella, EHEC (*Escherichia Coli*) and Hepatitis A. (*www.who.int/gho/phe/water_sanitation/burden_text/en/index.html*).

10. In addition to diarrhoea, the other main causes of water-related BoD are trachoma (20%) and intestinal nematode infections (6.5%).

11. Excluding the additional costs of deeper groundwater abstraction. The Texas State Water Plan 2012 does not estimate the environmental cost (in-situ values) of groundwater depletion, except for the Ogallala aquifer in Regions O and A.

References

Anderson, D. et al. (2000), "Estimated Annual Economic Impacts from Harmful Algal Blooms (HABs) in the United States", *Woods Hole Oceanographic Institution Technical Report*, *www.whoi.edu/fileserver.do?id=24159&pt=10&p=19132*.

Ara et al. (2006), *Measuring the Effects of Lake Erie Water Quality in Spatial Hedonic Price Models*, Ohio State University, presented at Environmental and Resource Economists, 3rd World Congress, Kyoto, Japan, 3-7 July.

Atech (2000), "Cost of Algal Blooms", *Report to Land and Water Resources Research and Development Corporation*, Canberra, ACT 2601, ISBN 0 642, *http://npsi.gov.au/files/products/riverlandscapes/pr990308/pr990308.pdf*.

Badiani, R. and K. Jessoe (2011), Electricity Subsidies for Agriculture: Evaluating the Impact and Persistence of These Subsidies in India, 31 March 31, *http://econ.ucsd.edu/CEE/papers/Jessoe_4april.pdf*.

Bommelaer, O.and J. Devaux (2011), *Coûts des principales pollutions agricoles de l'eau*, Série Economie et Évaluation, No. 52, Commissariat Général du Développement Durable, Paris.

Brink, C. et al. (2011), "Cost and Benefits of Nitrogen in the Environment", in *The European Nitrogen Assessment*, Sutton et al., Cambridge University Press, Cambridge.

Brody, S.D. et al. (2007), "The Rising Costs of Floods", *Journal of the American Planning Association*, Vol. 73, No. 3.

Burke et al. (2002), *Reefs at Risk in Southeast Asia*, World Resources Institute, *www.wri.org/publication/reefs-risk-southeast-asia 11/01/2012*.

Cutler, D. and G. Miller (2005), "The Role of Public Health Improvements in Developing Countries: The Twentieth-century United States", *Demography*, Vol. 42.

Dilley, M. and D. Hikisch (2009), "Water and Business. Claiming the Global Risk of Water Insecurity as a Strategic Opportunity for your Company", *www.awarenessintoaction.com/whitepapers/Water-Business.html*.

Dornbusch, D. and S. Barrager (1973), *Benefit of Water Pollution Control on Property Values*, US EPA 600/5-73-005.

Durand, P. et al. (2011), "Nitrogen Processes in Aquatic Ecosystems", in *The European Nitrogen Assessment*, Sutton et al., Cambridge University Press, Cambridge.

Esrey, S.A. (1996), "Water, Waste and Well-Being: A Multi-Country Study", *American Journal of Epidemiology*, 143(6).

European Environment Agency (EEA) (2005), "Source Apportionment of Nitrogen and Phosphorus Inputs into the Aquatic Environment", *EEA Report*, N° 7/2005, EEA, Copenhagen.

European Environment Agency (EEA) (2003), "Europe's Water: An Indicator-Based Assessment", 2003, *Topic Report*, 1/2003, EEA, Copenhagen.

Fewtrell, L. et al. (2005), "Water, Sanitation and Hygiene Interventions to Reduce Diarrhoea in Less Developed Countries: A Systematic Review and Meta-Analysis", *The Lancet Infectious Diseases*, 5:42-52 January 2005, Research into Environment and Health, University of Wales, Aberystwyth, UK.

Freeman, A.M. (1993), "The Measurement of Environmental and Resource Values: Theory and Methods", *Resources for the Future Press*, Washington, DC.

Gagnon (2007), "Health Costs of Inaction With Respect to Water Pollution in OECD Countries", ENV/EPOC/WPNEP(2007)9/FINAL.

Gleick, P. and L. Nash (1991), The Societal and Environmental Costs of the Continuing California Drought, Pacific Institute, Oakland, USA.

Grey, D. and D. Garrick (2012), "Water Security as a 21st Century Challenge", *Brief No. 1*, International Conference on Water Security, Risk and Society, University of Oxford, 16-18 April, January 2012, *www.eci.ox.ac.uk/watersecurity/downloads/briefs/1-grey-garrick-2012.pdf*.

Grizzetti et al. (2011), "Nitrogen as a Threat to European Water Quality", in *The European Nitrogen Assessment*, Sutton et al., Cambridge University Press, Cambridge.

Hertel, O. et al. (2002), "Assessment of the Atmospheric Nitrogen and Sulphur Inputs into the North Sea using a Lagrangian Model", *Physics and Chemistry of the Earth*, B, 27.

Hutton et al. (2008), *Economic Impacts of Sanitation in Southeast Asia*, Water and Sanitation Programme, World Bank, Washington, DC.

Hutton, G. and L. Haller (2004), *Evaluation of the Costs and Benefits of Water and Sanitation Improvements at the Global Level*, WHO, Geneva.

International Energy Agency (IEA) (2012), *World Energy Outlook 2012*, IEA, Paris.

Johnstone, N. and Y. Serret (eds.) (2006), *The Distributional Effects of Environmental Policy*, Edward Elgar Publishing, UK, *http://dx.doi.org/10/1787/9789264066137-en*.

Johnstone, N. and Y. Serret (Undated), "Determinants of Bottled and Purified Water Consumption: Results Based on an OECD Survey", *Internal OECD document*.

Kemper, K. and S. Foster (2004), *Economic Instruments for Groundwater Management*, GW-MATE, World Bank, Briefing Note Series, Note 7.

Kjellström, T. et al. (2006), "Air and Water Pollution: Burden and Strategies for Control", in *Disease Control Priorities in Developing Countries*, 2nd edition, The World Bank, Washington, DC.

Krysel et al. (2003), *Lakeshore Property Values and Water Quality: Evidence from Property Sales in the Mississippi Headwaters Region*, Bemidji State University.

Leake S.A. (2004), "Land Subsidence From Ground Water Pumping", *U.S. Geological Survey*, U.S. Department of Interior.

Margat, J. (2008), *Les eaux souterraines dans le monde*, BRGM-UNESCO, Paris.

Maria, A. (2003), *The Costs of Water Pollution in India*, CERNA, Paris, France.

Michael, H. et al. (1996), "Water Quality Affects Property Prices: A Case Study of Selected Maine Lakes", *Miscellaneous Report*, 398, Maine Agricultural and Forest Experiment Station, University of Maine.

Nagaraj, N. et al. (2003), "Sustainability and Equity Implication of Groundwater Depletion in Hard Rock Areas of Karnataka – An Economic Analysis", *Indian Journal of Agricultural Economics*, 2003(3).

National Research Council (NRC) (1997), *Valuing Groundwater: Economic Concepts and Approaches*, Committee on Valuing Groundwater, NRC, Washington, DC.

Nayak, S. (2009), "Distributional Inequality and Groundwater Depletion: An Analysis Across Major States in India", *Indian Journal of Agricultural Economics*, 2009(1).

Nogawa, K. and T. Kido (1993), "Biological Monitoring of Cadmium Exposure in Itai-itai Disease Epidemiology", *International Archives of Occupational and Environmental Health*, 1993(63).

OECD (2012), *OECD Environmental Outlook to 2050: The Consequences of Inaction*, OECD Publishing, *http://dx.doi.org/10.1787/9789264122246-en*.

OECD (2011), *Greening Household Behaviour: The Role of Public Policy*, OECD Publishing, *http://dx.doi.org/10.1787/9789264096875-en*.

OECD (2008a), *Costs of Inaction on Key Environmental Challenges*. OECD Publishing, *http://dx.doi.org/10.1787/9789264045828-en*.

OECD (2008b), *OECD Environmental Outlook to 2030*, OECD Publishing, *http://dx.doi.org/10.1787/9789264040519-en*.

OECD (2005), *OECD Reviews of Health Systems: Mexico 2005*, OECD Publishing, *http://dx.doi.org/10.1787/9789264008939-en*.

Olmstead, S.M. (2010), "The Economics of Water Quality", *Review of Environmental Economics and Policy*, Vol. 4, No. 1.

Peters, E. (2003), *Propagation of Drought through Groundwater Systems*, Dissertation, Wageningen Universiteit.

Pimentel, D. (2005), "Environmental and Economic Costs of the Application of Pesticides Primarily in The United States", *Environment, Development and Sustainability*, Vol. 7.

Prüss et al. (2002), "Estimating the Burden of Disease from Water, Sanitation and Hygiene at a Global Level", *Environmental Health Perspectives* 10(5), WHO Geneva.

Prüss-Üstün, A. et al. (2004), "Comparative Quantification of Health Risks, Global and Regional Burden of Disease Attributable to Selected Major Risk Factors", Mezzatti et al. (eds.), WHO, Geneva.

Reddy, R. (2003), *Cost of Resource Degradation Externalities: A Study of Groundwater Depletion in Andhra Pradesh*, Center for Economic and Social Studies.

Reyes, H. et al. (1998), "La mortalidad por enfermedad diarreica en México: ¿problema de acceso o de calidad de atención?", *www.scielosp.org/pdf/spm/v40n4/Y0400403.pdf*, 4 January 2012.

Ruta, G. (2005), "Deep Wells and Shallow Savings: The Economic Aspect of Groundwater Depletion in MENA Countries", Background Paper to *Making the Most of Scarcity: Accountability for Better Water Management Results in the Middle East and North Africa*, World Bank, Washington, DC.

Sarkar, A. (2011), "Socio-economic Implications of Depleting Groundwater Resource in Punjab: A Comparative Analysis of Different Irrigation Systems", *Economic and Political Weekly*, 12 February.

Sathyapalan, J. and S. Iyengar (undated), "Economic Valuation of Ecosystem Changes Due to Salinity Ingress and Ground Water Depletion: A Study of Coastal Gujarat", Center for Social Studies, Veer Narmad South Gujarat University Campus, Surat, India.

Shah, T. (2007), "The Groundwater Economy of South Asia: An Assessment of Size, Significance and Socioecological Impacts", in *The Agricultural Groundwater Revolution: Opportunities and Threats to Development*, M. Giordano and K.G. Villholth, CAB International.

Shah, T et al. (2007), *Groundwater: A Global Assessment of Scale and Significance*, International Water Management Institute (IWMI), Sri Lanka.

Shah, T. and C. Scott (2004), "Groundwater Overdraft Reduction through Agricultural Energy Policy: Insights from India and Mexico", *Water Resources Development*, 2004(2).

Soares, R. (2007), "Health and the Evolution of Welfare across Brazilian Municipalities", *Journal of Development Economics*, Vol. 84.

Terrel, B. et al. (2002), "Ogallala Aquifer Depletion: Economic Impact on the Texas High Plains", *Water Policy* (4)2002.

Tsur, Y. (1990), "The Stabilization Role of Groundwater when Surface Water Supplies are Uncertain: The Implications of Groundwater Development", *Water Resources Research*, Vol. 26, 1990(5).

Tudela, F. (2005), *The Costs of Inaction with Respect to Human Health Impacts from Pollution*, presented at the EPOC High-Level Special Session on the Costs of Inaction, 14 April 2005, Paris, OECD.

Warren, J.P. et al. (1974), *Cost of Land Subsidence Due to Groundwater Withdrawal*, Texas Water Resources Institute.

World Health Organization (WHO) (2008), *The Global Burden of Disease, 2004 Update*, WHO, Geneva, *www.who.int/healthinfo/global_burden_disease/GBD_report_2004update_full.pdf*.

World Health Organization (WHO) (2002), *Water and Health in Europe*, A Joint Report from the European Environment Agency and the WHO Regional Office for Europe (ed. Jamie Bertram).

WHO and United Nations Children's Fund (UNICEF) (2013), *Progress on Sanitation and Drinking-water – 2013 Update*, WHO Press, Geneva, *www.wssinfo.org/fileadmin/user_upload/resources/JMPreport2013.pdf*.

WHO-UNICEF Joint Monitoring Programme for Water Supply and Sanitation (JMP) (2008), *Progress on Drinking Water and Sanitation, Special Focus on Sanitation*, UNICEF, New York and WHO, Geneva.

World Bank (2007), *The Cost of Pollution in China*, Economic Estimates of Physical Damages, Washington, DC.

World Bank (2004), *Economic Instruments for Groundwater Management*, GW-MATE Briefing Note Series, Note 7.

World Economic Forum (2013), *Global Risks 2013*, An Initiative of the Risk Response Network, 8th edition, Geneva.

ORGANISATION FOR ECONOMIC CO-OPERATION AND DEVELOPMENT

The OECD is a unique forum where governments work together to address the economic, social and environmental challenges of globalisation. The OECD is also at the forefront of efforts to understand and to help governments respond to new developments and concerns, such as corporate governance, the information economy and the challenges of an ageing population. The Organisation provides a setting where governments can compare policy experiences, seek answers to common problems, identify good practice and work to co-ordinate domestic and international policies.

The OECD member countries are: Australia, Austria, Belgium, Canada, Chile, the Czech Republic, Denmark, Estonia, Finland, France, Germany, Greece, Hungary, Iceland, Ireland, Israel, Italy, Japan, Korea, Luxembourg, Mexico, the Netherlands, New Zealand, Norway, Poland, Portugal, the Slovak Republic, Slovenia, Spain, Sweden, Switzerland, Turkey, the United Kingdom and the United States. The European Union takes part in the work of the OECD.

OECD Publishing disseminates widely the results of the Organisation's statistics gathering and research on economic, social and environmental issues, as well as the conventions, guidelines and standards agreed by its members.

OECD PUBLISHING, 2, rue André-Pascal, 75775 PARIS CEDEX 16

(97 2013 11 1 P) ISBN 978-92-64-20239-9 – No. 60791   2013

IWA Publishing's authorised EU representative for General Product
Safety Regulations is Diane D'Arras, 15 rue Duret, 75116 Paris,
France, e-mail: safety@iwap.co.uk.

Printed and bound by CPI Group (UK) Ltd, Croydon, CR0 4YY

12/05/2026

02108886-0013